Today is a Good Day!

Attitudes for Achieving Project Success

Today is a Good Day!

Attitudes for Achieving Project Success

by Alfonso Bucero

First Edition

Multi-Media Publications Inc.

Oshawa, Ontario

Today is a Good Day! Attitudes for Achieving Project Success
by Alfonso Bucero

Managing Editor: Kevin Aguanno
Copy Editor: Peggy LeTrent
Typesetting: Peggy LeTrent
Cover Design: Troy O'Brien

Published by:
Multi-Media Publications Inc.
Box 58043, Rosslynn RPO
Oshawa, ON, Canada, L1J 8L6

http://www.mmpubs.com/

All rights reserved. No part of this book may be reproduced or transmitted in any form or by any means, electronic or mechanical, including photocopying, recording or by any information storage and retrieval system, without written permission from the publisher, except for the inclusion of brief quotations in a review.

Copyright © 2010 by Multi-Media Publications Inc.

Paperback   ISBN-13: 9781554890569
Adobe PDF ebook   ISBN-13: 9781554890576

Published in Canada and printed simultaneously in the United States of America and the United Kingdom.

CIP data available from the publisher.

Table of Contents

Acknowledgements 7

CHAPTER 1
Your Attitude 9

CHAPTER 2
How to Attract Project Success 23

CHAPTER 3
Make Your Plan for Success 39

CHAPTER 4
Make a Commitment 57

CHAPTER 5
Convert Your Project Issues into
Opportunities 69

CHAPTER 6
Your Words Make a Difference 81

CHAPTER 7
How Are You? 99

CHAPTER 8
Don't Complain About Your Projects 109

Chapter 9
  Associate With Positive Professionals 123
Chapter 10
  Grow Through Your Fears 133
Chapter 11
  Get Out There and Fail 145
Chapter 12
  Networking 161
Chapter 13
  Conclusions 177

About the Author 181

Acknowledgements

In May 2000, I had the opportunity to visit New Delhi, India, where I attended a project management conference as a keynote speaker. It was a real challenge for me because my English level was not very good at the time. However, I applied my courage attending that conference and communicated not only through my words, but from my heart. I learned a lot in that conference about the great power of a positive attitude. Since then, I started to apply the three Ps (Passion, Persistence and Patience) as the three principles of my personal and professional life. It still works for me.

Later, in May 2006, I attended the PMI Leadership Program, being part of the Leadership Master Class. As soon as I graduated, I made a commitment to write this book, telling my experiences with stories about the value of a positive attitude and the difference it makes for the Project Manager.

This book is dedicated to different sets of project stakeholders:

- All Project Managers and executives who contributed to this book with their opinions, experiences, and practices to this project, directly or indirectly.

Today is a Good Day!

- Laura and Olin Jennings from "The Jennings Group", who taught me the principles of leadership, helping me gain more and more insight about my passion.
- My best friend and better project professional, Randall L. Englund who helped me edit this book, shared with me many metaphors that inspired my passion for project management and encouraged me to continue writing.
- My wife Rose and my children, who gave me unwavering support and encouragement during the process of writing this book.
- My mother, who at the age of 84 helped me to understand the huge damage negativism may do.
- The memory of my father, who taught me how much you need other people.
- Dr. David I. Cleland, who read my manuscript and gave me great ideas and suggestions during the writing process.

I thank you all very much because you helped me and encouraged me to get my, *"Today is a Good Day"* project finished.

CHAPTER 1

Your Attitude

If you don't like something, change it. If you can't change it, change your attitude. Don't complain.
—Maya Angelou, US author & poet (1928)

Introduction

Early in my career I had a negative attitude regarding my job and towards the projects I managed. That negative *disposition* generated more problems than advantages. I created a negative image of myself in front of my colleagues, team members and managers. The result was not good. I transmitted negativism to my managers and team members.

The maturing process for me, led me to change my thinking. I needed an attitude check! By changing my attitude, I changed my world. This is such a fundamental, life changing experience, that I now feel compelled to share it with you, my readers.

A Definition of Attitude

The dictionary defines attitude as, *a position of the body or manner of carrying oneself, a state of mind or a feeling; disposition, an arrogant or hostile state of mind or disposition.* Attitude

Today is a Good Day!

is the preference of an individual or organization towards or away from things, events or people. It is the spirit and perspective from which an individual, group or organization approaches community development. Your attitude shapes all your decisions and actions. Attitude is very difficult to define with precision as it consists of qualities and beliefs that are non-tangible. We are used to talking about the attitude of individuals, but it is important to recognize that project teams and organizations also have attitude. Usually, however, when we talk about an organization's attitude, we use the term "organizational culture". When we talk about project's attitude, we use the term project culture. The Project Manager's attitude dramatically affects team attitude.

For instance, an important team attitude is confidence. The development of a project presents tremendous challenges to a project team. Sometimes it can even feel like an act of faith. An enormous amount of detail is collected, analyzed, organized, and assimilated into a functional *"whole"*. On very large efforts, only a few key individuals may possess the total *"big picture"*, and this may be at varying levels of completeness. This ambiguity can from time to time test the confidence of the project team members. Given these uncertainties, how does a team feel assured and confident of success throughout the process, and have this reflected in individual team member attitudes?

The following figure shows key qualities and beliefs that, from my experience, determine whether or not an individual, team or organization, has the attitude needed to successfully lead or actively participate in a project.

Your Attitude

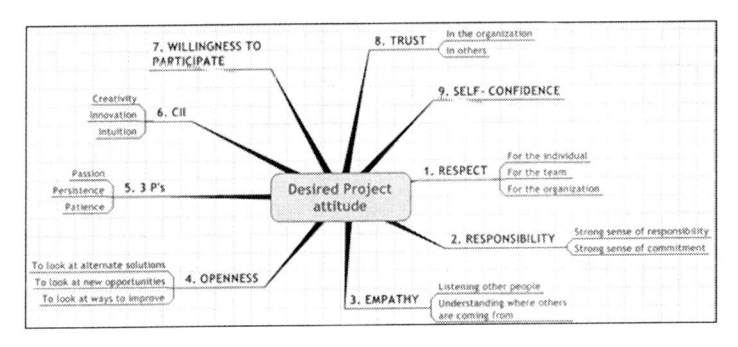

[Diagram: Desired Project attitude with branches:
1. RESPECT — For the individual, For the team, For the organization
2. RESPONSIBILITY — Strong sense of responsibility, Strong sense of commitment
3. EMPATHY — Listening other people, Understanding where others are coming from
4. OPENNESS — To look at alternate solutions, To look at new opportunities, To look at ways to improve
5. 3 P's — Passion, Persistence, Patience
6. CII — Creativity, Innovation, Intuition
7. WILLINGNESS TO PARTICIPATE
8. TRUST — In the organization, In others
9. SELF-CONFIDENCE]

William James said, "*The greatest discovery of my generation is that human beings can alter their lives by altering their attitude of mind*". I strongly believe attitude is a choice. The average Project Manager wants to wait for someone else to motivate him or her. The Project Manager perceives that his or her circumstances are responsible for the way he or she thinks. But which comes first, the attitude or the circumstances? The truth is that it does not matter which comes first. No matter what happened to you yesterday in your project or in your organization, your attitude is your choice today. Your attitude determines your actions.

Attitudes are a secret power working twenty four hours a day, for good or bad. Attitude is a brain filter through which you experience the world. Some people see the world through the mind of optimism while others see life through a mind of pessimism. I obviously found some people in the middle—not very optimistic but also not very pessimistic.

Here is an example to explain the difference between a positive attitude and a negative attitude regarding a project: Imagine a manager asking a Project Manager, "*How is your project going?*" The Project Manager answers, "*It is going bad, as always*". With that approach your enemies will be happy and your friends will be sad. A better answer would be, "*We are progressing. I have detected that some project activities are delayed, I also detected some project issues, but my team is taking corrective actions and everything will be under control very soon. I'll keep you informed about project progress".*

People with a positive attitude focus on project solutions. People with a negative attitude focus on problems and issues. Project Managers with a negative attitude dramatically affect project success. It is the attitude of the Project Manager to the project and to the team that will determine the attitude of the project to the Project Manager. We shape our own projects. We have the choice of choosing the attitude that will make a success of our projects.

Attitude is a great reflection of your spirit. Look at yourself. Are you happy as a Project Manager? Be honest. The environment you find around yourself is a mirror of your attitude. If you have a problem then you should start asking questions. Your team will change when you change for them. My advice is to treat those around you as would like them to treat you. Everyone needs recognition, gratitude and a kind word. My experience is that the attitude you start with often has a marked influence upon the final outcome of any venture. Good attitudes are often the introduction to an opportunity and also the final arbiter to success.

A key desire of professionals is to be respected and appreciated. How can we make that happen? Instead of thinking about what's missing, count your blessings. Don't see project limitations, identify risks and see opportunities. I have managed many projects outside my city of residence, staying away from Monday to Friday, far from home. I needed to be positive with my people, in spite of the strain on me personally. I always believe I should lead by example. When I talk to project management audiences about Project Manager attitude, I use pictures, jokes and video clips to help people understand and remember what I said. I demonstrate an attitude that I care about communicating effectively with the audience and will use various means to make my message as clear as possible. Make your attitude to project stakeholders *"fresh"*—open your *"project window"* to get fresh air and keep people moving forward through energy.

The Concept of Project Window

Your attitude is your window to project success. We all start out in life with a good attitude. Just watch our children. They are always laughing and giggling. They have a positive disposition and they love to explore new things. My youngest child is eleven-years-old. When he learned to ride a bicycle, he fell down many times, but he was always laughing. He fell down and tried again, and again. He spent some days learning to ride the bicycle, and he had a lot of fun doing that activity. I observed how he never complained; his objective was to learn.

Most activities we do in a project fail in the beginning. We must try again and again until we get it; we must be *persistent*. We experience a lot of criticism in project environments. Our project window gets splattered by bad comments and feedback from colleagues, executives and sometimes by project sponsors. *"You cannot always control*

your project issues but you can control your own thoughts". The problem is, the dirt keeps building up, and too many people do nothing about it. I had a dirty window when I worked for a multinational company as a Project Manager over many years. And the longer I stayed in that job, the filthier my window got. I saw no possibilities of moving forward; I needed fresh air. How could I? My window was splattered with a lot of negativity. Management feedback was too negative ... at least, that was my perception. There was a poor project management attitude in that organization. When I asked for feedback from other colleagues regarding my perception, they reinforced my opinion. That encouraged me to move forward.

Wash Your Project Window

Fortunately, I discovered that all I had to do was clean off my window and look outside. I had to improve my attitude so I could see the professional world clearly again. After I removed the grime from my window, a whole new world opened up for me. My frustration lifted and I had more confidence. For the first time in many years, I could see the magnificent possibilities what my profession, as a project management practitioner, had to offer. I was much more conscious of my lack of knowledge. The opportunities for learning in the project management field became huge.

Then, I was able to do a career move and do work that I absolutely love. Now, "*Today is a good day*" is my favorite sentence. I founded my company six years ago. I passed through problems, difficulties, issues, but now I feel free as a bird, and happy doing what I want to do. I started my "project" with a positive attitude, not forgetting that there is no project without problems. We must deal with problems and issues in our life, and then in the projects we manage. My attitude has changed completely and this reflects positively in my professional and personal life. It affects the projects I manage, it affects my project stakeholders, and obviously it has a big positive business impact. Furthermore, it affects my family, my most critical project. I am happy and I try to transmit positiveness to everyone most of the time. Not every day I am able to achieve it 100%, but I try to learn from my experience day by day. I believe my work is enjoyable and I try to make it enjoyable for my team every day.

Take Charge of Your Attitude

Now your job will be to keep your project window clean. My objective in this book is to provide a little encouragement. And other people can encourage you, too. But you must do it; nobody else can do it for you.

You always have a choice. You can leave the filth on your window and look at your projects through a smeared glass. But there are consequences to that approach and they are not very pretty. You will go through projects with a negative attitude and feel frustrated. You will be unhappy. You will achieve only a fraction of what you are capable of achieving.

There is a better way. When you choose to take out your squeegee and clean your window, life will be brighter and sunnier. You will be healthier and happier. You will set some ambitious goals and begin to achieve them. Your dreams will come alive again. You are probably thinking that's easy to say but difficult to do. Granted, some really devastating things may have happened to you. You may have endured much suffering. Perhaps you are going through some tough times right now. But, even under the worst circumstances, I contend that you have the power to choose your attitude. I am not saying it is easy. But I believe the choice is yours.

Let me tell you about my personal experience managing a project. Some years ago I managed a project in the South of Spain. At the beginning of the project my team

Today is a Good Day!

was not there. I was alone in front of the customer because my team was finishing another project. For almost a month I was alone with my customer and showing a lot of enthusiasm regarding the project. I worked with my customer on the strategic part of the project and spent time analyzing various project stakeholders and talking about the commitment needed from everyone for project success. I organized presentations and workshops to create teamwork until my project team was on the customer site. I felt very stressed by that situation, but I never showed my stress in front of the customer, trying to demonstrate self control. I exercised for one hour every day and I tried to keep myself healthy. My self-discipline was critical in that situation because many times the project environment is not healthy.

It was a very stressful situation. I believe that a happy Project Manager is not a professional in a particular set of circumstances but rather a competent professional with a set of attitudes.

Attitude and Project Success

Let's say you clean off your window and develop a positive attitude. You are smiling. You sit home and think positive thoughts. Will that alone lead you to outrageous success and the realization of your fondest dreams? No, it won't because there's more to success than just having a great attitude.

To maximize your potential and achieve your goals, my advice is to apply certain principles of success that have helped me achieve very good results. I hope they will serve you in the same way. In the next chapters of this book, I take you through these principles, step by step. You will learn about confronting your project fears, overcoming adversity, harnessing the power of commitment and more. Still, you may be wondering, what do these success principles have to do with attitude?

Your Attitude

TODAY IS A GOOD DAY. Without a positive attitude, you cannot activate the other principles. However, the promise that attitude is everything is hollow. In fact, if you believe that attitude is everything, it may actually hurt you more than help you. There is another factor that stands in the way: talent. Attitude is the difference maker. Attitude is not everything but is one thing that can make a difference in your projects when talent is present. The aim of this book is to show you that attitude is the difference maker in your projects. Then, I always apply two rules every day:

- Rule # 1: Today is a Good Day!
- Rule # 2: If today is not a Good Day for you, please apply rule #1

There is not a single part of your current life that is not affected by your attitude. Your future will definitely be influenced by the attitude you carry with you from today forward. When you combine a positive attitude with the other success principles, you become unstoppable!

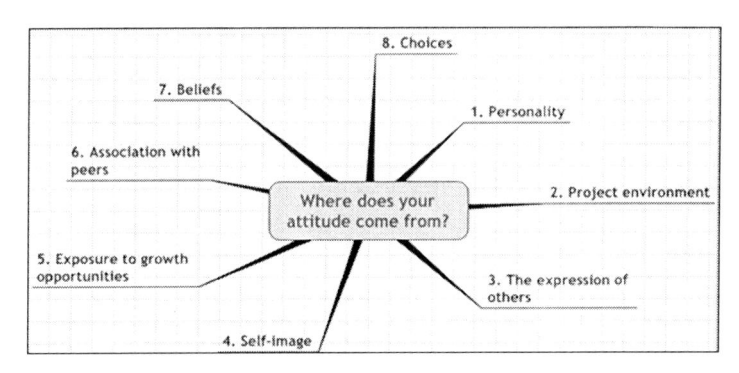

Where does your attitude come from?

1. Personality
2. Project environment
3. The expression of others
4. Self-image
5. Exposure to growth opportunities
6. Association with peers
7. Beliefs
8. Choices

If attitude is so important, then you may be asking yourself, where does attitude come from?

1. **Personality:** Who you are? Your personality type impacts your attitude. That is not to say that you are trapped by your personality, because you are not. But your attitude is impacted by it.
2. **Project Environment:** What's around you? The environment you are exposed to in your projects has an impact on your attitude. Do you have enough management support from your upper manager in the project you manage? Do you have a sponsor assigned for your project? Do you have any project constraints, deadlines? It may be hard to predict exactly what will happen to a person's attitude based on his or her early environment, but you can be certain that it made an impact of some kind. Clearly, environment makes a difference.
3. **The expression of others:** What you feel. Most people can remember the harsh words of a parent, coach or teacher, years or decades after the fact. Many times the hurts that cause people to overreact to others come as the result of negative words from others. Likewise, positive words can have an impact on a person's attitude.
4. **Self-image:** How you see yourself has a tremendous impact on your attitude. Poor self-image and poor attitudes often walk hand in hand. It is hard to see anything in the world as positive if you see yourself as negative. If you do not change your inward feelings about yourself, you will be unable to change your outward actions toward others.
5. **Exposure to growth opportunities:** Players must accept the cards dealt to them. However, once they have those cards in hand, they can choose how

they will play them. They decide what risks and actions to take.

6. **Association with peers:** It is an anthropological observation that you become like the people you spend a lot of time with.

7. **Beliefs:** What forms and sustains your attitude are your thoughts. Try to know others better. Every thought you have shapes your life. What you think about your neighbor is your attitude toward that person. The way you think about your project is your attitude toward that project. The sum of all your thoughts, filters, and assumptions comprise your overall beliefs.

8. **Choices**: Most people want to change the world to improve their lives, but the world they need to change first is the one inside them. That is a choice. You do not choose your project, your personality type, or your genetic make up. Everything you are and nearly everything you do is not up to you. You must live with the conditions you find yourself in. However, what you decide is how you do things. The longer you live, the more choices you make and the more responsible you are for how your life is turning out.

To change your life and to be more successful as a Project Manager, make a choice to take responsibility for your attitude and to do everything you can to make it work for you. Your attitude truly can become a difference maker. It is up to you. As in the following case study, I needed to make a choice; it was not easy to do but I did it.

Today is a Good Day!

Case Study

I managed an IT infrastructure project in the North of Spain. The project involved two third parties and a team of twenty people. It was not a very complex project initially; however, it was a strategic project for the customer (Eusko Jaurlaritza) and also a very important project for the multinational company I represented. Furthermore, I was assigned as a Project Manager for this project just two months after joining the company. Everyone in that organization was looking to me. It was the first project I managed there and that represented a big challenge for me.

I had six previous years of experience as a Project Manager with other companies. It was not very difficult for me to begin the project and deal with people on the customer site. However, one month into the project assignment, my father was diagnosed with lung cancer. I was managing the project outside my city of residence, and I spent the whole week away from home. I felt very sad and stressed because of that situation, but I needed to make my choice. My first option was to resign from the project; but the consequences would be bad because everyone had high expectations about my job, the customer was happy with my work, and my team members felt loved by me. The second option was to continue managing the project but also to communicate the problem to everybody. To make my choice was not easy. My father was dying slowly. I remembered his words spoken during my childhood. He said that ownership and commitment were so important for feeling good as a human being and also as a professional. He taught me always to separate people from the problems. Every time I was with him he asked me about my work, insisting, "*Please do not worry about me*". Then I made my choice—I continued managing the project. Every weekend I was with my father at the hospital taking care of him. He never complained; he gave me love and positivism all the time. The project was moving forward. I had to manage third party problems, technical issues, people problems—nothing foreign to me as a Project Manager. Six months later when my father died, my manager asked me why I did not tell him anything about my situation before. I never thought that I would be able to put up with my stress. I could not change my circumstances, but I could choose my attitude.

Few people in my organization understood my attitude. I always say that we are not the center of the Universe. We live in an environment that continuously changes. We can not change that environment but we can change and adapt our attitude regarding our project environment. In that project I could not allow people to worry about my personal problem. I needed to be the leader and to lead by example. I needed to smile frequently to my people who were away from their homes and families too. They were looking at and observing my behavior, and so I concluded that my choice was the only possible option.

I do not like to be considered a hero; it was my decision at that time. I refused any public reward from my company because I felt responsible for my decision.

Many readers of this book have their own horror stories to tell. The lessons learned here are that we as Project Managers are responsible for choosing our attitude. As the poet Maya Angelou says, *"If you don't like something, change it. If you can't change it, change your attitude. Don't complain"*.

Summary

Attitude is an attribute that influences almost every action we might take as Project Manager. No matter how wealthy we are, how well trained, how successful, whatever our station in life, we will finally be judged on the human element of attitude that we project. This chapter covers the concept of attitude. I say, *"Attitudes are a secret power working twenty four hours a day, for good or bad. Attitude is a brain filter through which you experience the world. Some people see the world through the mind of optimism while others see life through a mind of pessimism".*

The main ideas of this chapter are:

- Your attitude is your window to project success. We all start out in life with a good attitude.
- Wash your project window.
- Take charge of your attitude by keeping your project window clean
- TODAY IS A GOOD DAY. Without a positive attitude, you cannot activate the other principles. However, the promise that attitude is everything is hollow. It serves as a difference maker.

This book addresses emotional issues, to help the project professionals cope with their leadership role, and to learn how to connect emotionally with the real payoffs that exist. To manage others successfully, you must manage yourself. Before going to the Chapter 2, take the time to complete the self-assessment in the Appendix in this book. By doing so, you will end with an action plan that will help you face the challenges ahead with success and integrity. Make a plan to get what you want and start today.

CHAPTER 2

How to Attract Project Success

Success is not the result of spontaneous combustion. You must set yourself on fire.

—Reggie Lech

We Become What we Think About

I strongly believe that *we become what we think about*. I believe that our thoughts determine our actions. The Project Manager must be a believer from the beginning to the end of the project. If the Project Manager does not believe in the project, he or she will not be able to convince team members.

Robert Collier offered this insight: "*There is nothing on earth you can not have – once you have mentally accepted the fact that you can have it*". A corollary thought especially relevant in the project management world is that *you can have anything you want, you just can't have everything.*

When I speak at seminars I use a very popular Spanish joke that expresses how important beliefs are for Project Managers. A Spanish man meets a gypsy man, and the gypsy man says, "*I have a horse to sell you*". The Spanish man says, "*A

Today is a Good Day!

horse, for what? I do not need your horse." The gypsy man says, "Oh! You need it. It is a wonderful horse. It wakes up early in the morning and prepares breakfast for the whole family. It's a great horse". The Spanish man says, "Ok man, I don't believe you, but I'll buy the horse". Then the Spanish man buys the horse. Two months later the two men meet again, and the Spanish man says to the gypsy man, "You were lying. It's an awful horse. It's crying all night. It's disturbing my neighbors. I don't want it". The gypsy man answers him, "You know man, continue talking about the horse that way and you will not be able to resell it".

We as Project Managers must sell our projects in organizations. First we must start selling the horse ourselves. If we are not able to believe in our horse, we will not be able to sell it. When you constantly think about a particular goal, then you will take steps to move toward that goal.

Let's say we have a Project Manager who thinks he will be able to convince a customer about the benefit of a solution. Like a human magnet, that Project Manager will attract people to influence the customer and convince them about the benefit of that solution.

The idea that we become what we think about has also been expressed as the Law of Dominant Thought. This means there is a power within each of us that propels us in the direction of our current dominant thoughts. The key word here is dominant. We have an internal power within each of us that propels us in the direction of our current thoughts. Obviously a little positive thinking does not produce positive results. You need to practice it and discipline yourself about thinking positively every day until it becomes a habit.

The way you live your project is sending messages to your team members. You are preparing them for project failure or success. It is up to you!

Adjust Your Attitude

Let me share a personal example to demonstrate the power of our thoughts.

In 1992, when I was a Project Manager at HP Spain, I presented my first paper in English at an HP Project Management Conference in the US. I was not fluent at speaking English at the time. However, I was passionate enough to write and submit a paper in English. My dream was to be able to present in English on my professional subject. I wrote a case study about the customer project I was managing at the time in Madrid, Spain. The conference organizers answered me by asking for some changes and I did my best. I prepared my first English presentation. Many later I went to San Jose, California, and presented my paper to a worldwide HP audience. I was very nervous at the beginning of my presentation because of my level of English. However, I was getting better as I went through my presentation. Why? Because I believed in that project and I had lived that experience with my team. At the time my English speaking ability was not important. What was most important was my passion in telling my story to the audience.

That presentation changed my behavior. Since then I have been sending papers to international Congresses and Symposiums, sharing my experiences with different audiences every year, forcing myself to write and speak better and in English. My mentor is my good friend and professional, Randall L. Englund, who helped me to progress professionally and personally and who helped me to write better in English. Some time later, after that first paper was submitted, he told me that acceptance was borderline; what swung the decision towards acceptance was the authentic message that came through—he was able to sense an attitude and story that needed to be shared. Today, fifteen years later I am a *PM Network* magazine Contributing Editor. I am the author of a book in Spanish with Mr. Englund. A book on *Project Sponsorship* (Jossey-Bass, 2006).

Project Sponsorship was written with only three face to face meetings. Mr. Englund lives in California and I live in Madrid, Spain. We exchanged many emails while writing our book and took advantage of international project management conferences to review and agree on book material. Once our project was finished and we were very happy about our deliverable. One main lesson learned was enthusiasm in the project and believing in the possibility to make a change and make a contribution along with just doing it.

My dreams have been *converted to reality* because my thoughts were oriented to achieve a clear goal. This excellent experience has been wonderful for me. It was priceless because it taught me that you can achieve your goal when you believe in yourself and keep your thoughts focused on the positive. You then attract others who want to work together with you. Furthermore, your organization benefits from this situation too.

Attitude and Action

I have been writing about the importance of positive thinking and positive thoughts. However, I have not written about where action fits into this process. We can not achieve any results without actions. But thoughts and ideas precede actions and when our thoughts are positive, immediately some actions come to mind. I know from personal experience that when I have a positive attitude, I feel compelled to take action and nothing can stop me. It does not matter if I make mistakes because I know I will be able to amend them.

I believe that is why a positive belief system is the starting point for the achievement of any goal. When your thoughts focus on achieving your goal, you begin taking the necessary actions to move forward. Let me give an example. When I wrote my first book, I was working as a Project Manager in a multinational organization. Sometimes I had long journeys, and I started to write after dinner which I considered as a daily task. Some days I was too tired and I could not write a word. Other days I spent two or three hours writing down my thoughts. I always needed to see something tangible and accomplished little by little. I printed out my work and that act encouraged me to continue until I completed my book.

I believe managing projects in organizations work with a similar approach. The Project Manager and his or her team need to achieve small deliverables as soon as possible because it generates motivation among the team members.

Your Circumstances

We are not always conscious of creating a lot of negativism every day in our environment. Your beliefs brought you to where you are in your career today, and your thinking from this point forward will take you to where you will be in the future.

Many years ago, I desired to be in the project management field. I began my career as a programmer, then a software engineer, and then team leader until I achieved my first goal which was to be a Project Manager. I was lucky because the organizations I was working for gave me possibilities for professional development. The activities I participated in over a period of years came about because of my dream to be a Project Manager, and I ultimately achieved that goal. My desire of becoming a Project Manager moved me to ask Project Managers about what they did and to find out more information about the profession. That helped me to develop my professional development plan and my personal vision. It was the way I found PMI (Project Management Institute) and discovered the huge amount of opportunities the project management profession had. It has kept me in touch with more than 700 worldwide project professionals over the last fifteen years.

I know there are some people in whom these trends are innate and other people in whom they are not. However, it is possible to change your attitude about your career and also about your life. I believe Project Managers must be ready to serve their team members, customers, and a variety of stakeholders.

Change Your Thinking

What are you thinking about yourself every day? Do you have positive or negative thoughts? Each of us has an internal voice. Many times what we say is negative, critical and self-limiting. We create our own obstacles, nobody else does it. Perhaps you find yourself thinking "I can not do this" or "I always mess things up". These thoughts work against you. Instead, repeat to yourself that you can and will accomplish your goal, your project or your activity. Many times we are not taking into account the words we use on a regular basis. We continually use negative ideas to express ourselves – Don't turn right. Don't think negatively. These statements make our point consciously, but how are these innocent messages being received at the unconscious level?

Every time you use a negative your brain is first registering the exact opposite of what you mean. Let me give you an example. We met a friend the other day with a broken toe and asked her what happened. She replied, *"I was shopping and had just picked up a tin of paint when a salesperson said, 'Careful, don't drop it on your foot'. A few moments later, the paint slipped out of my hands"*. The unconscious processed the helpful salesperson's message as, *"Drop it on your foot – Don't,"* and she followed the initial command to the letter!

Think of how often children are disciplined with negatives. Don't fall. Don't touch that. Don't ... Don't ... Don't. Sometimes what may seem like disobeying could just be their unconscious processing the experience of falling or touching. Have you ever thought, *"I don't want to fail," "I don't want to be in this job all my life," "I must not miss this sale?"* Think about it. By the time your brain has got to the *"don't"* or *"must not"*, your whole body has already received the message, *"fail, stay in the job, and miss the sale"*.

So if you want to think positively, the key question to ask yourself is, *"If I don't want "X" what do I want in its place?"*

Answering this question precisely gives your brain something to work toward. The more you can specify what you will see, hear and feel the more the brain is given a map to create new solutions and guide it to its outcome. So, if you find yourself thinking something like, *"I don't want to be nervous in front of a group,"* ask yourself, *"what would I be doing instead?"* You might answer, *"I will have steady hands, breathe easily, feel relaxed in my stomach, make eye contact, have a clear voice, be humorous, etc."*. The more detail and the more specific you are, the better.

In order for you to be effective, your brain not only needs a specific goal, it also needs a DIRECTION to achieve that goal. If it does not have a direction it wanders randomly, avoiding what you don't want, but not going anywhere in particular. You can aim in a specified direction and maintain that course, or you can say, 'Good luck' – and indeed, whatever will be will be–but it will be outside of your control.

One way to get used to moving in a desired direction is to continually ask yourself, *"What do I have to do NEXT to achieve my goal?"* As far as the unconscious mind is concerned, the goal is like a map. The direction is like a compass. You need both to navigate successfully. Strategies for motivation use the mind's ability to move you toward what you want, and move you away from what you don't want.

My recommendation is to use positive words and comments about yourself as a Project Manager and your goals.

Here are suggestions to help you become more positive and get the results you want:

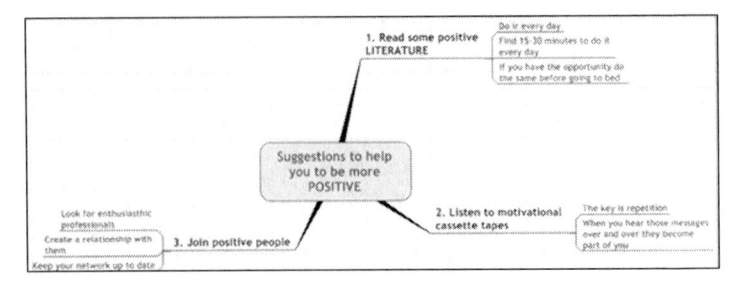

```
                              Do it every day
            1. Read some positive   Find 15-30 minutes to do it
            LITERATURE               every day
                                     If you have the opportunity do
                                     the same before going to bed

         Suggestions to help
          you to be more
             POSITIVE

Look for enthusiastic              2. Listen to motivational    The key is repetition
professionals                         cassette tapes            When you hear those messages
Create a relationship with   3. Join positive people            over and over they become
them                                                            part of you
Keep your network up to date
```

- Read some positive literature every day. Find 15-30 minutes in the morning to do this task. If you have the opportunity do the same before going to bed.

- Listen to motivational cassette tapes every day. The key is repetition. When you hear these messages over and over they become part of you – and you begin to implement them to improve your life

- Join positive people. Look for positive and enthusiastic professionals and establish a professional relationship with them.

If you follow these suggestions, it will make a phenomenal difference in your life. I promise you these techniques work if you have the discipline to stick with them. My lesson learned is: change your thinking and you will change your results in your activities, your projects and your life.

Overnight Project Success

Positive thinking is a process that takes time and patience; it is not an overnight success. Positive project thinking requires effort, commitment and patience. On the other hand, positive thinking does not mean absence of problems. You will have a lot of setbacks along the way. However, if you continue to believe in yourself, being persistent, you can overcome those obstacles.

Believe me, you are moving constantly in the direction of your dominant thoughts. Everything you achieve in your career flows from your own thoughts and beliefs. Negative thinking yields negative results and positive thinking produces positive results.

It makes no sense to think negative thoughts, unless you want to get negative results. Then, from this point forward, choose your thoughts wisely and use these experiences to get fantastic results in your career and your projects.

These ideas regarding optimism may help you throughout a project life cycle:

- **Calm-down** – don't expect everything is working your way. Life is a wheel and whenever you are at the wrong side of the turn, chill-out. Go out and find outlets to release pressure and negative feelings. Forcing the issues won't likely work.

- **Downplay but do not take for granted** – if things go wrong, downplay them to ease the negative emotions. If you were not able to join an outing or important gathering because of prior non-enjoyable work, just assure yourself of the benefits you will get from work and the gathering isn't going to be that fun as you thought it would be.

- **It's not the people** – I believe that people should not be blamed when they make mistakes. Processes and events should be factored at the top of the priority. Do not blame yourself when something bad happens. Look at events and other external factors that could have caused the incident. If you were not able to accomplish something, do not tell yourself you are not good enough; instead, recognize you need for more training and education to deliver at high levels. Give yourself the benefit of the doubt and be thankful that you are a living being getting exposed to a variety of experiences.
- **Stand-up** – don't be afraid to fail and stand back up. Optimistic people learn and use their failure to get back on track. Reflect on what's happening and how to learn from it.
- **Same feathers flock together** – make sure you hang out with optimistic people. Their attitude will help you stay positive and optimistic.

CASE STUDY: Creating a PMO

In the Solution Business, communication and documentation with the client and within the delivery organization are critical to the project delivery process. The Project Manager needs to focus on managing customer expectations to get things done. To maintain this focus, he or she needs to be relieved of many internal organizational concerns.

The difficulty increases when a project culture does not exist—or what is there is very weak—in an organization that delivers customer projects, and the management team only focuses on numbers and results. They ask for good project results but do not worry about how to establish and create the right environment for project success.

In this kind of organization, upper managers were not supporting the Project Manager, and Project Managers were on their own when dealing with internal and external stakeholders during the project life cycle. Sponsorship was an unknown term in this environment.

A Project Office implies some kind of innovation because it changes the way to proceed—creating the ability for the Project Manager to keep focused on the client and perform high quality project management. All internal and external stakeholders and their expectations must be analysed, the team must be assigned, and all the activities need to be divided in functional groups and most importantly, a very effective and empowered team must be created.

Background

The foundation for HP Spanish Project Office began in September 1999. As the organization grew in terms of projects and people, to know more about project status became a real issue from the management's perspective. Then the management team decided to implement a Project Office to relieve Project Managers of low value activities. They asked me to work on a "solution proposal". This proposal was presented, studied, discussed and finally accepted by the management team in February 2000. The PMO project started on March 1st, 2000 at the Madrid office.

The Spanish Project Office arose from the need to relieve Project Managers of administrative tasks associated with managing a project. The Hewlett-Packard Consulting (HPC) Organization believed the Project Office should help the Project Manager improve efficiency, facilitate getting the

Today is a Good Day!

right tools, and align services with the needs of the project environment.

Generally, at a corporate level, project offices are regarded as project management centers of expertise. HP decided that the professionals who staff these project offices should be experienced and trained in project management skills. At the local level, we use the Project Office to change our culture from a reactive style to a project oriented organization. This culture shift largely occurs by demonstrating or modeling the new behavior. Our approach is to get results on a subset of projects initially, and use these successes as models for others to adopt our process. It was a step by step process, achieving small tangible things happening in the organization.

Mission and Objectives

It took some weeks to come to an agreement with the management team about the reason for this particular project. Why do we need a Project Office?

I explained to the management team that Project Office adds value to project team members, providing mentors, consultants, training, tools to be more effective and SIC (structured intellectual capital). I was always enthusiastic and positive with my comments, talks and discussions.

The Project Office adds value to the organization, providing culture shift to project management, reusable tools and techniques, document and methodology support, global recognition, profitability improvement and quality support. The Project Office also adds value to our customers, providing a visible sign of the HP commitment, competent HP team support and more effective and quick answers.

Critical Success Factors

In all the projects I managed in my professional career, I found that project success depends on *how well you work with and lead people*. That behavior may attract people to achieve project success. I learned that the project office approach must be aligned with the culture of the organization. Technical problems can be solved with new releases or different hardware or software, but it is different when we talk about people interactions and relationship among team members. Although we identified some factors as critical in the Project Management Office (PMO) implementation project, one of the most important critical is to focus on being prepared to answer questions and demands from different project stakeholders. Each consultant and Project Manager expects the PMO is there to help them on a daily basis and that means to be prepared for a world of uncertainty. Many times the type of demand is driven by pressure in terms of time or expectations, and we as the PMO need to transmit feasibility and securely. *I always asked for proactive behavior from each PMO team member. I discovered* the following success factors:

- Scope agreement and clear setting expectations between all users and stakeholders (it took some weeks of meetings and validations).

- Forming, storming and norming the PMO team (80% of my team was formed by subcontracted people. I employed time for initial training, methods, tools and procedures).

- Clear defined functions, roles and responsibilities for the PMO (I talked one on one and verified each person's expectations).

- Sponsorship from the upper level management (I asked the general manager to ask people to use the project office services).

- Clear communication plan deployment.
- Communicate project status periodically to the management team and to the end users.
- Positive attitude, team service orientation.

Lessons Learned

I learned a lot from the PMO project. One the most essential behaviors for me was transmitting passion and positive behaviors to my team members and executives.

- A PMO team needs time for:
    - *Forming:* It takes a lot of effort and time. I need to invoke patience many times.
    - *Storming:* Allowing people to generate and brainstorm ideas and thoughts.
    - *Norming:* Discipline and guidelines are necessary for all type of people but are mandatory for inexperienced team members.
    - *Performing:* I always gave to my team the benefit of the doubt.
- People will support a project office (and communications in general) when they see its value and how it links directly to positive business impact. Achieving fast short results was key for project success.
- Finding and provoking energy in your team members is really important to achieve project success. Empowering my team members was necessary to manage team and project issues.

Summary

I believe that our thoughts determine our actions. The Project Manager must be a believer from the beginning to the end of the project. We, as Project Managers, must sell our projects in organizations, especially if you are a Project Manager who thinks you will be able to convince a customer about the benefit of a solution. Like a human magnet, you will attract people to influence the customer to become convinced about the benefit of that solution.

Please remember these key ideas about positive thinking:

- My experience is that when I have a positive attitude, I feel compelled to take action and nothing can stop me. It does not matter if I make some mistakes, because I know I will be able to find and implement a solution.

- We are sometimes not conscious that we create a lot of negativism daily in our environment by our comments and reactions, and people always observe you as a leader.

- My recommendation is to use positive words and comments about yourself as a Project Manager and your goals.

- Positive thinking is a process that takes time and patience; it is not an overnight success. Positive project thinking requires effort, commitment and patience.

So, how can you attract Project Success? Before moving to chapter three, go to the Appendix and do your self-assessment on chapter two. Start today.

Today is a Good Day!

CHAPTER 3

Make Your Plan for Success

You must first clearly see a thing in your mind before you can do it.
—Alex Morrison

Since the age of twenty six I dreamed of being a Project Manager. I started to work in the Information Technology field as a programmer, then as an analyst and then as a project leader, leaving and joining several organizations. As many readers can attest to as well, I worked very hard to earn every bit of my actual position. Now I have my own firm and work on my passion, project management. I believe you must dream before achieving things because when you dream, unconsciously you generate synergy for the steps to follow that achieve your objectives.

Word class-swimmers also incorporate the power of imagery to reinforce in their mind exactly how they want to perform. Many top competitors mentally envision a successful outcome before actually achieving it in the "real" world.

Visualization refers to the practice of seeking to affect the outer world via changing one's thoughts. Creative Visualization is the basic technique underlying positive thinking. The concept arose in the United States with the

Today is a Good Day!

nineteenth century New Thought movement. One of the first who practiced the technique of creative visualization was Wallace Wattles (1860–1911)[2] who wrote "*The Science of Getting Rich*", which he believed was based on spiritual principles.

Creative visualization is the technique of using your imagination to create what you want in your life. It is purported to be accomplished by various means including expecting the best, focusing intensely on a desired goal, and repeating affirmations and the use of vision boards.

Creative Visualization is distinguished from normal daydreaming in that Creative Visualization is done in the first person and the present tense – as if the visualized scene were unfolding all around you; whereas normal daydreaming is done in the third person and the future tense – the "you" of the daydream is a puppet which the real "you" is watching from afar. Visualisation practices are a common form of spiritual exercise, especially in esoteric traditions. In Vajrayana Buddhism for example, complex visualisations are used to attain Buddhahood.

Visualization, however, is not something reserved solely for athletes or movie stars. In fact, it is something you have used since childhood to create the circumstances of your own life. Visualization is often described as *"movies and pictures of the mind," "inner pictures"* or *"images"*. We all store pictures in our minds about the type of relationships we deserve, the degree of success we will attain at work, the extent of our leadership ability, and so on.

We begin to develop "inner pictures" early in life. If we were criticized, or felt unworthy, as youngsters, we recorded the events (and the feelings associated with those events), as images in our minds. Because we frequently dwell on these pictures (both consciously and subconsciously), we tend to create life situations that correspond to the original image. For example, you may still hold a vibrant image of being criticized by a manager in a project meeting. I still recall a manager I worked for as a Project Manager in a multinational company; he always gave me negative feedback. He never gave me a positive word. He reprimanded me in front of my team members. He made me feel very bad and decreased my self confidence. The picture remains in your mind even when you are not always conscious about it. These pictures exert tremendous influence over present actions or project activities.

Take Ownership and Create Your Vision

Not all mental pictures can be traced to your past. You are constantly generating mental movies and/or pictures based on your relationships, project experiences and other events. No matter what the source of your mental images, there is one point that I want to drive home: you, and only you, are in control of your own movies.

Let's try a short experiment. Think about a glass of orange juice. Does that create a picture or image for you?

Now think about a horse. Can you see it? Change the color of the horse to pink. In a fraction of a second, you formed an image of the pink horse. Can you bring back the picture of the glass of orange juice? Of course you can.

You have control over the pictures that occupy your mind. However, when you don't consciously decide which pictures to play, your mind will look into the *"archives"* and keep re-playing old movies on file in your mental library.

Then think about your vision as a Project Manager, analze how things are happening in your projects and try to visualize future images of desired outcomes. Think about how you would like to be treated by your team members and visualize the results. It is true that you can not change many situations but you definitely have the freedom to choose the way in which you can imagine them. Those mental images dramatically affect the future of projects as well as relationships with team members and other project stakeholders.

Change the Meaning of the Old Picture

It does not serve you to deny what happened in a past experience, no matter how painful or disappointing. For instance, you cannot change the fact that you were criticized by the manager. But you can alter your interpretation of the event. When I told my story about being criticized by my manager in front of my team, my perception was very negative. But I finally understood it some years later.

At the time of an original criticism, the meaning assigned to the experience is often "I am not good enough" or "my opinions are worthless". While this may be the interpretation of an inexperienced professional, the thought inadvertently gets carried into your life as a Project Manager. But today, you can consciously choose to view the situation differently. For example: The manager may have disagreed with you, but it was not a statement about your intelligence or your overall worth as a person. The manager may have been in a bad mood because of something else or may have had a point of view to share that eluded me at the time. That is the positive vision I am practicing now, recognizing the opportunity I had to learn through the experience with that manager. A powerful statement to enact is to say, *"I can choose to think differently about this situation"*.

I strongly believe, that many times we make incorrect interpretations of the words and actions coming from other people. Sometimes, it is because of our education, society, environment or culture. That is related to different people's motivational factors. Different people have different perceptions according their motivational values, according Dr. Elias Porter.

Create New Pictures

We can create new mental pictures whenever we choose to do so. And when we develop (and concentrate on) new images that evoke powerful feelings and sensations, we will act in ways that support those new pictures. So, the first step is to create an image of your desired outcome. You are limited only by your imagination.

Some Project Managers are terrified over public speaking. In survey after survey, it is listed as the first fear that people have in organizations. So, when most people are asked to consider making a speech, what kinds of pictures do they run through their minds? They see themselves standing nervously in front of the audience. Perhaps they have trouble remembering what they want to say. Run these images over and over on your mental screen and you can be sure that you won't have much success as a speaker.

Instead, form a picture in your mind in which you are confidently giving your presentation. The audience members are listening to your every word. You look sharp. Your delivery is smooth. You tell a funny story and the audience laughs. At the end, you get a warm round of applause. People come up afterward to congratulate you. Do you see how these kinds of mental images can help you become a better speaker?

However, recognize that the pictures in your mind are not fulfilled overnight. But, by being patient and by persistently focusing on these mental images, you will automatically start acting in ways that support your vision.

At the beginning of my professional career, I was involved in a project acting as a project leader by accident (without any training and knowledge). I had a team of six

Today is a Good Day!

people, all of them older than me. My instinct told me I needed to listen to them before acting. Then, I started the project plan process with all of them. When the draft plan was ready, I was asked to deliver a presentation at the project kick-off meeting. It was my first presentation in my professional life in front of sixteen people. I was very nervous and had butterflies in my stomach. But I listened to one of my oldest team members who told me, *"Alfonso, be quiet, before starting your presentation look to the persons in the room and imagine all of them are in pyjamas"*. I put that small trick into practice, and it relaxed me a lot. Immediately I told myself, *"Alfonso you are the person who knows more about this particular project. Go ahead and achieve a good result. You can do it"*.

The reader may think that is a joke. However it is absolutely true. That event was a key differentiator in my life and also a key learning experience as a professional. Since that time I have used pictures and visualizations for various purposes, and they work for me. Before presenting I rehearse, and rehearse, becoming more confident.

Picture Your Way to Success

You as a Project Manager or as an executive are involved in selling a product or service. It is vital that you see yourself as succeeding on a consistent basis. If you are not getting the results you want, there is no question that you are holding onto pictures of sales mediocrity, or sales disappointment as opposed to sales success.

Right now, think about your next meeting with a prospect. In your mind, how do you see the encounter? Are you confident and persuasive? Are you enthusiastically explaining the benefits of what you are offering? Is the prospect receptive and interested in what you are saying? Can you vividly see a successful outcome to your meeting?

Make Your Plan for Success

Remember that you are the producer, director, script writer, lighting coordinator, costume designer and casting director of your own mental movies. You get to choose how they turn out. By mentally rehearsing and running successful outcomes through your mind, you are paving the way for success in your sales career.

Of course, if you currently run images through your mind where the prospect rejects your ideas and has no interest in your presentation, you will attain very limited success from your sales efforts. You will attract those people and those situations that correspond to your negative images.

Relax and Involve Your Senses

What is the best method to use when concentrating on your new images? Your mind is most receptive to visualization when you are calm and not thinking about a lot of things simultaneously. So, sit down in a comfortable chair at home, close your eyes and do some deep breathing exercises to clear your mind and relax your body. Now, develop images that involve as many senses as you can. The more sights, sounds, smells, tastes and touches you put in your pictures, the more powerful the "pull" for you to make your vision a reality.

Here is an example. Let's say you have always dreamed of owning a beach-front house in the Caribbean. Picture the white and peach colored-house. See the green palm trees slowly swaying in the gentle breeze. Smell the salt air. Feel the warm sand between your toes. Feel the sunshine on your face. Is this not paradise? And all this can be yours if you hold onto this image and do what it takes to achieve it. Also, remember that those images associated with strong emotions have even more power, so be sure to add positive feelings to your vision. For instance, when visualizing your ideal job, combine the vivid mental picture and the physical senses with the terrific

Today is a Good Day!

emotions of pride and satisfaction you will have working in that new position.

Finally, do not be concerned with the quality of your images at the outset. Some people can create lively color pictures while others have trouble getting anything more than a fuzzy image. It is also possible you may only be able to get a particular feeling at the beginning as opposed to a clear image.

In any case, do not worry about it. Do the best you can and do not compare yourself to anyone else. There will always be someone better and others who are worse. Just be who you are. Your images will become sharper over time. The key is to spend several minutes each day running these new movies in your mind.

Do a Commitment to Yourself

Commitment is an interaction dominated by obligations. These obligations may be mutual, or self-imposed, or explicitly stated, or they may not. Distinction is often made between commitment as a member of an organization (such as a sporting team, a religion, or as an employee), and a personal commitment, which is often a pledge or promise to ones' self for personal growth.

It is very powerful to formulate images of successful outcomes and to run them through your mind. But there is another technique you can use to accelerate your success. You can create visual aids to move you toward what you want. Let me share with you my story, a story of passion, persistence and patience:

In 1996, after presenting some papers written in English (which is not my native language), I realized that writing was enjoyable for me and I set an objective of getting a book published within the next few years, at least in the Spanish language. I was travelling a lot for project purposes over the next two years, but as soon as the number of trips slowed down I started working on my first book. And I got it. My first book, *Dirección de proyectos – Una Nueva Visión* was published in Mexico by Lito Grapo Editors, thanks to a good colleague and friend from Mexico, Rodolfo Ambriz,PMP.

After that book was published, and because of my good relationship with Randall L. Englund, PM Executive consultant, speaker and author, I got an article published at PM Network from PMI. Randall helped me edit that article and encouraged me to continue writing. I followed his advice and set my new objective which was to be a column contributor of PM Network. And I did it one year later. Similar stories have been happening over the last five years and now I have written this book for you.

One of the tips I followed to achieve my objectives was to put that objective as an item in my daily tasks, as a reminder. That means to visualize the objectives. Make sure to look at your diary once a day and believe that you are moving toward that goal.

Getting the Job You Want

You can use visual reminders to your advantage in a lot of ways; it is not limited to commitments. Here is an example of one of my colleagues, who we will call Francis. Francis was presented as a candidate for the PMI, Barcelona Chapter Board of Directors elections. While Francis stands an excellent chance of winning this election (and realizing his dream of becoming a president), he is still a little nervous and doubts creep into his mind now and again. I suggested to Francis to make a handwritten sign where he will see it every day. I also recommended that he write these words on a card he could carry in his wallet.

Today is a Good Day!

By looking at those words throughout the day, Francis is conditioning his mind to view himself as a PMI Chapter Vice President. He is going to start thinking about sitting in the Board of Directors room table in a Board meeting. As these images become stronger and stronger, Francis will take those actions that will bring this picture into reality. He will campaign more. He will make sure his party is doing everything possible to get the voters out on Election Day. While Francis could have formed strong mental images without the use of the sign, it had much more power with the visual aid. The sign is a reminder for Francis to think about being a Vice President and to run successful images through his mind.

Of course, there are no guarantees that this will work for another person or that it will always work for you. But, once you try this for yourself, I think you will find that it is an incredibly powerful aid to help you get what you want.

What is the project you need to manage, what do you want to achieve? Whatever it is, create a visual aid and your mind will get to work to bring that picture into your life.

It Works Both Ways

Be very careful when using visual reminders. Some people use negative aids and with very serious consequences. Bumper stickers offer a prime example.

While riding in my car a few years ago, I noticed a bumper sticker on the car in front of me. The bumper sticker read, "I feel sad, I feel sad, so off to work I go". In the last few years, I have seen this same bumper sticker over and over.

There is nothing funny or harmless about this message. When you put something like that on your car, you are programming your mind to keep you sad. For example, every morning Rose steps outside to greet the day and sees the statement *"I feel sad"*. When it is time to leave work, she goes back to her car and sees *"I feel sad"*. This idea becomes embedded in her subconscious mind. She forms mental pictures associated with feeling sad.

If you ask Rose why she lost that project deal, she will say that she had bad luck. The truth is, Rose is careless about what enters into her mind. The "harmless" little bumper sticker of today becomes your reality tomorrow.

Rose is a perfect example of someone throwing more mud on an already dirty attitude window.

The Planning Process

If you don't plan, it doesn't work. If you do plan, it doesn't work either. Why to prepare a plan?

The nice thing about not planning is that failure comes as a complete surprise rather than being preceded by a period of worry and depression. The planning activities that you, with the help of your team members, will need to do for the project are listed below:

- To recruit and build the team.
- To organize the project.
- To identify and confirm the start and end dates through a project schedule.
- To create the project budget.
- To identify clearly the customer requirements for the final outcome.
- To define the project scope boundaries - what it included and not included in the project.
- To write a description of the final outcome.
- To decide who will do what.
- To assign accountability.

Today is a Good Day!

Planning for Success

To be successful, a project must have a plan that demonstrates what is possible, that shows an overall path and clear responsibilities, contains the details for estimating the people, money, time, equipment, and materials necessary to get the job done; and will be used to measure progress during the project and act as an early warning system.

Planning for success is also based on team member's commitment. An example of commitment was the "Manhattan project": It was, among other things, a gigantic industrial and engineering construction effort, run by the military under great secrecy, rapidly accomplished, using unorthodox means, and dealing in uncertain technologies. It's central purpose was to develop and build an atomic bomb as quickly as possible that could be used to end the war. It got underway in June 1942, but only with the appointment of Army Corps of Engineers officer Colonel (quickly promoted to Brigadier General) Leslie R. Groves on September 17, 1942 did it become an all-out crash program. In less than three years the first bomb was tested on July 16, 1945 in the New Mexico dessert. Three weeks later on August 6th the Japanese city of Hiroshima was bombed followed by the bombing of Nagasaki on August 9th. The war was over five days later on August 14th.

One of the characteristics of the Manhattan Project was the unconventional practice of conducting the *research, development, and production phases simultaneously* rather than following a step-by-step sequential path, that is the normal and slower way. Because every minute counted during wartime all the steps were compressed, done in parallel, rather than one after the other. Occasionally there was a problem but the wrong choice was soon righted. In one example, machines to make the gaseous diffusion barrier material, based on one design, were being installed in a newly built factory. With work just about completed General Groves decided that an alternative design would be better. And so the just installed machinery was stripped out of the plant and new machines were put in their place. This practice of compressing the different stages does not always work and should be a cautionary tale to any future projects.

Total program authority was vested in Groves. He had the complete support of the president and the other high officials of the administration. The full resources of the United States Treasury were available to him. In the end the Project cost about $2 billion, which would be about $30 billion in today's dollars. The project was initially understood to involve possible national survival against an evil enemy bent on world domination that may have been trying to build an atomic bomb of its own. The objective was clear, unmistakable, finite, and well defined. Compartmentalization, in addition to maintaining security, kept people focused on their assignments and responsibilities to achieve it. Each element had its own task, and all were carefully allocated, assigned, and supervised so that the sum of the parts resulted in the accomplishment of the mission.

Command channels were clear-cut, well understood, and direct. Authority was delegated with responsibility. Large staffs were avoided, especially in Groves· Washington office. People at the higher levels knew one another from past experiences and could quickly communicate to solve problems and make decisions. Written communication was kept to a minimum. Most business was done verbally by phone or face to face. Groves" decisions were not based on staff studies, committee reports, written opinions of consultants, or the like.

The Manhattan Project was administered very much according to the organizational model and practices of the Army Corps of Engineers, not surprising since Groves, the purest of specimens, was its head. The model emphasized decentralization but through clear lines of command to the

Today is a Good Day!

top. The success of the Manhattan Project I believe lays within the culture and organization of the Army Corps of Engineers and the high quality of the officers that ran it. They were used to big projects. Size did not phase them. Groves in his earlier capacity, just prior to being selected Manhattan chief, oversaw more than eight billion dollars worth of domestic army construction projects during the mobilization period from 1940 to 1942.

    Groves always projected an optimistic attitude, which inspired others. Morale could only be sustained if everyone thought that the thing could be done. If Groves showed any doubt, hesitation, or fear, it might have infected the others and undermined the project. Groves normally set completion dates that he was sure could not be met. Keep the bar high and people will work harder to jump over it. He set completion dates which he was sure could not be met. It was only in this way that he could be certain that every effort would be made and no one could think of easing up if he had too easy a schedule. *"Success is not a matter of luck"*, Groves said, *"but the result of mental and physical capacity, of endeavor, of determination, and in large measure, of competent management"*.

    Examining the Manhattan project and such collaborative efforts, we conclude:

- To achieve success, start with superb and gifted people.

- It is best that they produce something tangible as opposed to working on an abstraction or an idea.

- Young people are normally more energetic, confident, and curious and thus are more likely to work harder and longer.

- It is all the better if the undertaking is driven by moral purpose. Put this special population in an isolated spot without any distractions. Living in

Make Your Plan for Success

> Spartan conditions makes work the focus, with no distractions.

- This tendency to escape into the work may result in ignoring or not having the time to reflect on what is being produced.

- The cooperation of the many parts toward realizing the overall goal is essential. Ensure that those below have faith in their leaders, and make sure that the leaders have faith in those below.

In conclusion, while it is important to learn and apply the lessons of history, we must remain cautious about making to easy of analogies.

Case Study

In the year 1992, I delivered my first project management presentation in a multinational company conference in San Jose, California. I made my decision to submit a paper at a conference the Project Management Initiative of HP organized. It was a challenge for me, because my English speaking level was low.

However, I had a clear idea in my mind of what I wanted to share about my project experience with my colleagues of HP internationally. Then, I submitted my paper. Initially I got some comments and requests from the organizers and I did my best. Finally my paper was accepted.

I had three to four months to prepare my presentation, and as a project, I prepared my project plan. I found some HP Spanish colleagues who had worked in the United States, and I asked them for advice on how to review my plan and how to manage the project risks. I went to the HP library and I looked for past HP Conference proceedings, trying to get familiar with paper styles. And finally, I joined an English class

Today is a Good Day!

every day. During the first two months my English classes were focused on grammar and sentence construction and during the last month I rehearsed my presentation every day. My English teacher recorded me with a video camera, and we reviewed my expressions and sentences together (project snapshot). My first feeling watching myself in the video and listening to my voice was difficult. I did not feel good. However, I believe this was a great method to receive feedback and make improvements. I was encouraged by a Spanish colleague who lives in the United States and that reinforced my positive behavior.

I remember the faces of HP colleagues like Tom Kendrick, PMP, and other colleagues in that room. At the beginning I was very nervous but after some minutes I felt much better. I remembered how I practiced so many times to achieve my objective—delivering a presentation in English and being understandable by the audience at that specific event. I visualized my scenario many times, and I believed that I could do it.

And I did because my talk was successful. I was understood by almost everyone because of rehearsing many times in front of a mirror and watching the videos. When I saw my presentation recorded by a colleague I was quite surprised. My improvement through that long process was huge. My English was not perfect, but my delivery was good, showing my preparation and knowledge about the subject. So I was credible, and got some applause. I really believe that you can plan for your success, and if you are passionate, if you have persistence and also a lot of patience, you can achieve good results.

For some of you readers this example may not appear as a big success. It was not, but it was the first step in convincing myself that through planning, practice, and courage, I could accomplish my goal.

Summary

Visualization is often described as *"movies and pictures of the mind," "inner pictures" or "images"*. We all store pictures in our minds about the type of relationships we want, the degree of success we will attain at work, the extent of our leadership ability, and so on.

Not all mental pictures can be traced to your past. You are constantly generating mental movies and/or pictures based on your relationships, project experiences and other events. Then think about your vision as a Project Manager, analyze how things are happening in your projects and try to visualize your future image of the desired situations.

I strongly believe that many times we incorrectly interpret the words and actions coming from other people, perhaps because of education, society or culture and related to people's motivational factors. Public speaking, an often terrifying action, provides opportunities to change attitude through re-visualization and achieving a positive outcome.

You as a Project Manager or as an executive are involved in selling a product or service. It is vital to see yourself succeeding on a consistent basis. If you are not getting the results you want, you are probably holding onto pictures of sales mediocrity, or sales disappointment as opposed to sales success.

Be very careful when using visual reminders. Some people use negative aids and with very serious consequences.

It is very powerful to formulate images of successful outcomes and to run them through your mind. What is the project you need to manage, what do you want to achieve? Whatever it is, create a visual aid and your mind will get to work on bringing that picture into your life.

Today is a Good Day!

CHAPTER 4

Make a Commitment

*Courage is doing what you're afraid to do.
There can be no courage unless you're scared.*
—Eddy Rickenbacker
US WWI aviator & businessman
(1890 - 1973)

Courage, also known as bravery, will, intrepidity and fortitude, is the ability to confront fear, pain, risk/danger, uncertainty, or intimidation. After inventing the light bulb, Thomas Edison was asked where he drew inspiration from, and he said, *"I find out what the world needs, then I proceed to invent"*.

The key to getting what you want is the willingness to do whatever it takes to accomplish your objective. Now, before your mind jumps to conclusions, let me clarify that in saying *"whatever it takes"*. I exclude all actions which are illegal, unethical or which harm other people. So, exactly what do I mean by this "willingness"? It is a mental attitude which means that if it takes five steps to reach my goal, I'll take those five steps, but if it takes fifty five steps to reach my goal, I'll take those fifty five steps, and so on. We usually do not know how many steps will be required to reach our goal. This doesn't matter. To succeed, all that's necessary is that you make a commitment to do whatever it takes regardless of the number of steps involved.

Where does persistence fit in? Persistent action follows *commitment*. You must first be committed to something before you will persist to achieve it. Once you have made a commitment to achieve your goal, then you will follow through with relentless determination and action until you attain the desired result.

When you make a commitment and are willing to do whatever it takes, you begin to attract the people and circumstances necessary to accomplish your goal. For instance, once you devote yourself to becoming, say, a better Project Manager, you might suddenly *"bump into"* a new professional resource or "discover" a forum on this topic.

It's not as if these resources never existed before. It's just that your mind never focused on finding them. Once you commit yourself to something, you create a mental picture of what it would be like to achieve it. Then, your mind immediately goes to work, like a magnet, attracting events and circumstances that will help bring your picture into reality. It is important to realize, that this isn't an overnight process; you must be active and seize the opportunities as they appear.

The magic which flows from commitment has never been more eloquently or more accurately described than in the following words by W.H. Murray:

> *"Until one is committed there is hesitancy, the chance to draw back, always ineffectiveness. Concerning all acts of initiative there is one elementary truth, the ignorance of which kills countless ideas and splendid plans; that the moment one definitely commits oneself, then Providence moves, too. All sorts of things occur to help one that would otherwise never have*

occurred. A whole stream of events issues from the decision, raising in one's favor all manner of unforeseen incidents and meetings and material assistance, which no man could have dreamt would have come his way."

You do not have to know at the outset how to achieve your goal. Sure, you will be better off if you have a plan of attack, but it is not essential that every step be mapped out in advance. In fact, when you have the willingness to do whatever it takes, the right steps are often suddenly revealed to you. You will probably meet people you never could've planned to meet. Doors will unexpectedly open for you. It might seem like luck or good fortune is smiling on you; when in truth, you will have created these positive events by making a commitment and, thus, instructing your mind to look for them. Let me provide an example.

Last July 2008 I launched a new idea to the market, the Project Portfolio Event. It was a professional event, one day long on project portfolio management. I was committed in making that event happen. I had a positive attitude while dealing with that project. However, everything wasn't rosy in my path. Life was testing me to see how serious I was about achieving my aim (to make it happen). I found many obstacles. Customer registrations were only a few at the beginning, sponsors didn't pay us early, and the worldwide market crisis affected us very seriously. I made some mistakes and suffered disappointments and setbacks, some of which may have been quite severe and even may have tempted me to abandon my goal. For instance, I had five people on my team, and although I always encouraged them, when we only had three weeks left for the event, registrations were still low. But I did not give up, I talked to them and asked them to move forward. I continued to contact potential attendees and doubled the number of attendees two weeks before the event.

That's when it becomes important to follow the sage wisdom of Winston Churchill, who said, *"Never, never, never give up"*. If you have made a commitment to accomplish a goal, you can overcome temporary defeats, and you will triumph.

If you want to be an effective leader you must be committed. True commitment inspires and attracts people. It shows them that you have conviction.

Refuse to Quit

I have learned a lot about the magic of commitment from my ex-colleague and friend, John P. John developed his Project Management career at a computer manufacturer organization in the UK. Since a very young age, he liked Project Management unconsciously. Some years later he started to work at a multinational computer manufacturer company as a Project Manager. However, he had a clear project in his mind (retiring in his fifty's). Then his goal was how to save money through hard work doing his preferred work (Project Management). He managed many projects during his career, and he always tried to take on large challenges. He travelled a lot, and sometimes he felt very tired and exhausted. However, he never quit. He continued and managed his last program as program manager of the Y2K (year 2000) project for a multinational company. He is now in his 60's and happily retired. He is enjoying himself managing the project of building his new house. He is an example of persistence.

We are talking about commitment but we are also talking about persistence and patience, about keeping a good attitude in the face of rejection.

Another story is Michael, who lives in Madrid, Spain. At the age of 45, Michael had graduated from a Spanish Project Management school. Then he started to prepare to take a PMP exam and be certified.

On his first attempt at the examination he failed. On his second attempt, he again failed. And he also failed the third time, the fourth time, the fifth time, the sixth time, and the seventh time. His main problem was that he took the exam in English and English was not his strongest skill.

Most people would have quit, but not Michael. He was persistent, and he took the exam for the eighth time and he passed. This is a clear example about positive attitude and courage. Now he is managing and consulting projects, and he is happy doing that job. Refusing to quit is to show your courage. I am reminded of a quote from Miguel de Cervantes (Spanish author) that said, *"If you lose your wealth you lose a lot, if you lose a friend you loose much more, but if you loose your courage you lose everything"*.

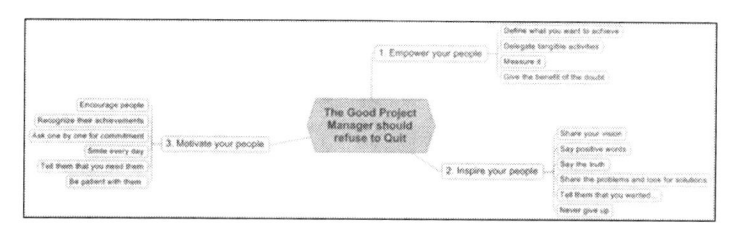

The good Project Manager must empower, inspire and motivate their team members to conquer project's daily obstacles and issues. Projects always generate problems. Work to solve them.

Time to Make a Commitment

Now, let's assume you have a goal in mind. The next question to ask yourself is, "Am I willing to do whatever it takes to achieve this goal?" If your answer is, *"I'll do just about anything, except that I won't do this and that..."*, then honestly, you are not committed.

Today is a Good Day!

And if you are not committed, it is likely that you will be derailed and not achieve your objective. For instance, many people start a new business with this approach, *"I'll give it six months to prosper. If things don't work out after six months, I'll quit"*. This isn't a mental attitude that leads to success.

When I left Hewlett-Packard in 2002 to start up my own business (a new project), many people from HP told me, *"Alfonso you are fool, business life is very difficult outside HP"*. However, I made my commitment to create my business. Now, almost seven years later, I am proud of being successful in my business. I passed over difficulties, people problems, customer issues, financial difficulties, but I got it and, I hope to continue managing my business for many years.

Now, I am not suggesting that you just bull ahead without a plan and hope for the best. Of course you should set timetables, deadlines and budgets so you stay on course and succeed as quickly as possible. But the reality is, despite your most careful plans, you don't know how long it will take to achieve your goal, and you can't foresee all the obstacles that will cross your path.

That's where commitment separates the winners from the losers. The committed person is going to hang in there and prevail no matter what. And if it takes a little longer than they thought, so be it. Those who are not committed are going to give up the ship when things don't go their way.

Now that you have learned about the power of commitment, it is time to apply the principle. So, go ahead and select a goal you have a burning desire to achieve. Make a commitment to do whatever it takes to achieve this goal. Start moving forward and get ready to notice and take advantage of all the opportunities that come your way. Then follow through with persistent action and get ready to succeed.

How to improve your commitment? Measure it, know what's worth dying for, and make your plans public. When I created my company I had to create new products and services. Most times I prepared a draft of the product description, content, objectives, audience, and I announced the availability date for that service or product to my customers and peers. That approach worked for me; it generated a strong commitment within me and with my people.

Commitment and Project Success

When I was assigned as a Project Manager for the General Treasury of Spanish Social Security in Spain, I never imagined what I would learn from that project about the power of commitment. Let me share my experience with you:

The Background

Hewlett-Packard and the Data Collection Center, of the General Treasury of the Spanish Social Security achieved one of the most advanced solutions for electronic document management. By using an image management system, they came up with a solution to digitice, store and process the Spanish Social Security contribution forms (TC1, TC2,..). Despite the tremendous volume of documents processed – about two million pages per month.

CENDAR, the Data Collection Control Center, is a Spanish public entity within the General Treasury for Social Security. It acts as a control center for information relating to the collection of contributions made by employees and companies as pertaining to Social Security.

Today is a Good Day!

The Situation

The General Treasury of Social Security used a manual data entry system to input information regarding Social Security contributions. Because of the substantial volume of paper generated by this operation, it became necessary to design a system that would allow for electronic storage of the forms for subsequent processing. In July 1991, the criteria for a public tender were published to purchase an information systems solution for the Electronic Data Collection Exchange Center (CENDAR). The proposal was won by HP.

The Solution

Good cooperation between the customer and HP in the CENDAR project team made it possible to make a clear definition and its objectives. The teamwork between the customer and the HP team and their commitment were the fundamental reasons for HP being able to meet the client's requirements on time, on cost and with the expected quality level. Long journeys were needed at the beginning of the project to create a good team. The team was formed by young people from 25 to 30 years old, but they were anxious and ready to learn. The cooperation during the solution design and implementation phases between CENDAR professionals and HP gave the customer a state-of-the-art document management system.

This real case showed that total commitment in respect of the client's requirements and prime responsibility for the whole project were the keys to success in development of the new system. The team I created and managed in this project taught me that together everyone achieves more. During this project I took care of creating a good environment, it was not easy because I had in my team people coming from different organizations, and then with different cultures and expectations. However you can see the main elements I consider for a good team environment:

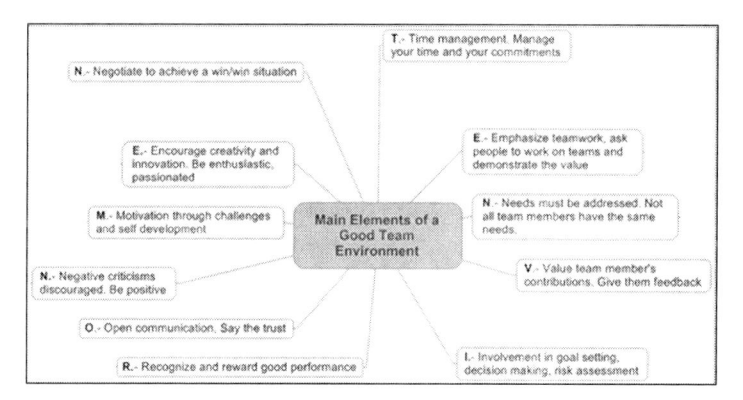

- **T.-** Time management. Manage your time and your commitments
- **N.-** Negotiate to achieve a win/win situation
- **E.-** Encourage creativity and innovation. Be enthusiastic, passionated
- **E.-** Emphasize teamwork, ask people to work on teams and demonstrate the value
- **M.-** Motivation through challenges and self development
- **N.-** Needs must be addressed. Not all team members have the same needs.
- **N.-** Negative criticisms discouraged. Be positive
- **V.-** Value team member's contributions. Give them feedback
- **O.-** Open communication. Say the trust
- **R.-** Recognize and reward good performance
- **I.-** Involvement in goal setting, decision making, risk assessment

Main Elements of a Good Team Environment

Case Study

When I finished my Computer Science Engineering degree in 1984, I started my PhD studies, but joined the Secoinsa firm as my first job. I started to travel frequently, so I interrupted my PhD studies some months later. However, I promised myself that some day I would finish my PhD. Although many of my professional colleagues told me: "you do not need a PhD now", I always thought if I pursue the possibility to discover something new, then I will learn something. Twenty five years later, at the age of 49, I made the commitment of achieving my PhD in Project Management. I did not find the Universities offering that specialty in Madrid for the weekends, so I made the decision to do it at Zaragoza University. Zaragoza is about three hours by train. But that was not a big deal for me.

Since the beginning of the program, the program Director transmitted negativism to all his students. I was very disappointed over his behavior, first of all because it is not a good skill to show up in a Project Manager practitioner. It impacted me very negatively. I almost decided to quit, but I was persistent, I was patient and determined to keep asking

Today is a Good Day!

myself the question, *"What do you want to achieve Alfonso?"*
I attended all the classes regularly during the first year to accomplish my objective. Unfortunately, for business reasons I could not finish some homework, and I failed three out of nine program subjects. But I did not quit and the next year I passed.

The third year I worked on my pre-thesis and presented my Thesis project. To stay focused on that purpose, I put my Thesis project on my agenda every day, in order to advance step by step. I am a frequent business traveller, and it was difficult at times. Then I got the DEA (Post Degree Studies Certificate) that is a pre-requisite to present the Thesis. I travelled to Zaragoza once a month to meet my mentor, who always told me, *"Alfonso, you will not be able to achieve it, because you have your own business, you must work for your customers to survive, you are managing a consulting firm, and it is not your first priority"*. Some not polite answers came to my mind. I always told him that I wanted to continue and I would be very persistent on that.

He still continues with the same talk, *"Alfonso you will not...., you will not be able to,..."* The last time I was very angry. I treated him to lunch and told him: "I love you too." I talked with him very clearly explaining that I could survive because of my passion, persistence and patience, even when some of my colleagues were quitting. Now our relationship has improved. He really is conscious about my objective. I am working on presenting a scientific article" to have published internationally. It will take some more months. I will be able to present my thesis by 2009. I will continue being persistent and enthusiastic about I can achieve what I want to. I will keep my passion, persistence and patience until the end of this real story.

Summary

The key to getting what you want is the willingness to do whatever it takes to accomplish your objective. Here are reminders of important best practices from this chapter:

- When you make a commitment and are willing to do whatever it takes, you begin to attract the people and circumstances necessary to accomplish your goal.

- If you want to be an effective leader you must be committed. True commitment inspires and attracts people. It shows them that you have conviction.

- The good Project Manager must empower, inspire and motivate team members to conquer daily obstacles and issues. Projects always generate problems; work to solve them.

- How to improve your commitment? Measure it, know what's worth dying for, and make your plans public.

- Always hang on to your passion, persistence and patience, not only in your projects, but also in your personal life.

Today is a Good Day!

CHAPTER 5

Convert Your Project Issues Into Opportunities

*Every adversity carries with the seed of
an equivalent or greater benefit*
—Napoleón Hill

When faced with problems or setbacks in the projects you manage, what is your immediate reaction? If you are like most people, your first impulse is to complain. *"Why did this have to happen to me? What am I going to do now? My plans are ruined."*

This response is only natural. However, after the initial disappointment wears off, you have a choice to make. You can either wallow in misery and dwell on the negative aspects of your project situation or you can find the benefit or lesson that the problem is offering.

Yes, you will probably face a period of uncertainty or struggle, but there's always a flip side to the difficulty. You see, a *"problem"* is often not a problem at all. It may actually be an opportunity. For instance, a problem may point out an adjustment you can make to improve certain conditions in your project. Without the problem, you would never have taken this positive action.

For example, you probably know or have heard about, someone who lost his or her job and then went on to start a successful business. Often, that person will tell you that if he or she had not been laid off, the new business would never have been started. What started as an adversity ended up as a golden opportunity. My colleague Randall Englund, shares how, during a low point in his career, he was compelled to find a new position. After an arduous search, he signed on to a Project Management initiative. That new position launched his career as a Project Management Consultant.

How about the times you were absolutely convinced that a particular job was perfect for you. You had a great interview and just could not wait for the offer. But the offer never came – someone else got the job. You were devastated. Days or weeks later a new job came along, and you realized that the first position was much less desirable than the one that came along later. The earlier rejection was, in fact, a blessing. Another example is the deal on the "dream house" which falls through only to be replaced by something even better.

The Benefit

Human beings come to appreciate the important things in life when we live through difficult times. Let me share with you a project I managed between 1990 and 1992 in the North of Spain.

I was assigned as a Project Manager for an infrastructure customer project. I had two people from HP assigned to my team, but I also had a big subcontractor as part of the team. That subcontractor had previous business relationships with my customer, so they were very confident in each other. I was the youngest among all main project

Convert Your Project Issues Into Opportunities

stakeholders, so I experienced lots of stress at the beginning of the project. I was in an uncomfortable environment. I was the Project Manager and the main person responsible for the project. This project was critical for HP, so I talked directly to my Project Sponsor (the HP Managing Director). I asked him for advice and support. He encouraged me to ask for help whenever I needed it. Then I tried to establish a closer relationship with the customer, talking to him, treating him to lunch, and providing him with technical explanations and project status. Little by little I gained credibility in his eyes. The subcontractor observed my behavior and treated me with lunch together. His approach started to become more open and sincere than at the beginning of the project. That achievement took more than six months, but I got it. My fear in front of the subcontractor disappeared and I demonstrated, first of all to me and then to others, that I could convert an issue into a benefit to facilitate project success.

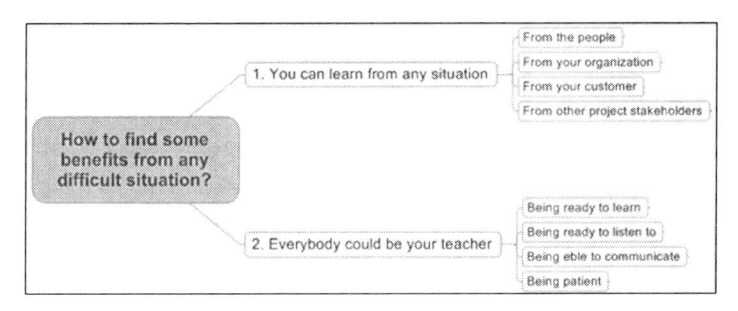

How to find some benefits from any difficult situation?

1. You can learn from any situation
   - From the people
   - From your organization
   - From your customer
   - From other project stakeholders

2. Everybody could be your teacher
   - Being ready to learn
   - Being ready to listen to
   - Being able to communicate
   - Being patient

I strongly believe that you can find some benefit from any difficult situation. You can learn from any situation and from anybody if you really think that everybody could be your teacher. In fact, you, the reader, could be my teacher.

Today is a Good Day!

From Project Failure to Project Success

The road to success often travels through adversity. In January 1990, I was assigned as a Project Manager for a $10 million customer project in the Spanish General Treasury of the Social Security. It was a software development and infrastructure project. Three months after the project began, the division which supported the software product we closed and fired all their employees without any advice. I felt extremely stressed in front of the customer. I felt there was a high probability that the project would fail. At the same time, one core project team member left the company. He was the most knowledgeable professional about key technical aspects on the project.

However, the company supported me and invested money to look for a subcontractor to provide us with an alternate software package. The company empowered and encouraged me to establish relationships with subcontractors. On one hand, I needed to find another professional with good technical knowledge. It was a great opportunity for me in terms of dealing with people from different organizations, with different cultures and different visions, and negotiating with them. At the same time, I had to interview various candidates to substitute for my project core team member. I learned a lot from that situation, personally and professionally. One year later the project finished successfully. The customer was happy using the system and my managers valued my effort and improvement.

My Project Management Career

My own professional career transition is another example of how benefits come from problems and difficulties. Adversity brings out our hidden potential. Since 1985 I wanted to be a Project Manager. I loved dealing with people and I enjoyed trying to manage issues and difficulties, but I probably wasn't aware of that.

Convert Your Project Issues Into Opportunities

In 1984, my first job was working for the Secoinsa firm, a Spanish computer hardware manufacturer, where I was a Technical Software Product Specialist. After one year of experience I had the opportunity to manage my first project with the *"Banco Hispano Americano"*. I managed that project by accident—without any project management experience. I had a team of four professionals, all of them older than me. But I was patient and perseverant trying to learn from them. It was the first time I told my team members, *"I need you"*. It worked so well. I ran into many difficulties within that one-year long project, I had to deal with people issues, customer and technical issues, but I survived. I learned to use common sense more and more. I became conscious of how difficult a project management discipline can be. On the other hand, I verified that it wasn't impossible.

I moved forward and one year later I moved to another company, Digital Equipment Corporation. I started in the Pre-sales Department as a technical consultant, but after a few months I was assigned to different projects, first as a team member and some years later as a technical team leader. I had the opportunity to receive basic project management training, and I discovered I liked it. After four and a half years I moved to ICL, another computer manufacturer, as a Project Manager.

I started acting as a Project Manager for ICL for projects in Madrid and Valencia. I developed leadership skills there and created good relationships with customers. However, the company situation and the lack of management support encouraged me to move on to my next job at HP Spain in the Madrid office.

I continued working as Project Manager at HP for almost fourteen years. It was a personal and professional challenge. I had the opportunity to learn and manage complex projects. I had a great opportunity to learn from my managers and Project Manager colleagues. I helped to start up three project offices. I discovered the existence of professional project management associations, and joined PMI (the Project Management Institute) in March 1993. A vast window opened for me when I attended my first International Project Management Congress. From my first attendance at a PMI Congress, project management began to become my passion. I discovered how much I had to learn and that that particular profession motivated me a lot. I took care of my professional network year by year, and by June 2002 I published my first book, *La Dirección de proyectos una Nueva Visión* which was published in Mexico with the help of my friend and project professional, Rodolfo Ambriz.

At the same time, I started teaching project management in Business Schools in Málaga and Madrid. It was a special effort but also a way of feeling professionally more recognized. On April 1, 2002 I left HP to start my own business. It has been a very difficult project, but I am happy now and I can say "Today is Good Day" after almost seven years. Now I have a team working with me that also believes in my principles: Passion, Persistence, and Patience.

During those seven years I had financial problems, people problems, and organizational problems. But I always had a positive attitude to learn from successes and failures. I always had the strong support of my wife Rose, who believes in me and taught me a lot about dealing with difficult people.

I continued writing articles, and submitting papers to Project Management Congresses. I wrote another book with my good friend Randall L. Englund on *Project Sponsorship* (Jossey-Bass, 2006), and I continue discovering how much I can learn from the project management field, from my customers, students, and colleagues. Expansion of the project management profession is a great challenge. The project management career will never end for me. The problems and

Convert Your Project Issues Into Opportunities

issues that come up when managing projects always represent opportunities to learn.

Project Adversity

Let's examine some ways in which adversity can serve you. Adversity provides you with perspective. Once you have recovered from a life-threatening illness, a flat tire or a leaky roof it does not seem so troubling any more. You are able to rise above the petty annoyances of daily living and focus attention on the truly important things in your life.

Adversity teaches you to be grateful. Through problems and difficulties, especially those which involve loss or deprivation, you develop a deeper appreciation for many aspects of your life. It is trite but true that you don't usually appreciate something until it is taken away from you. When you have no hot water, you suddenly value hot water. Not until you are sick do you cherish good health. The wise person continues to dwell on blessings, even after the period of loss or deprivation has passed. Remember, you are always moving in the direction of your dominant thoughts; therefore, concentrating on what you have to be grateful for brings even more good things into your life.

Adversity brings out your hidden potential. After surviving a difficult deal or overcoming an obstacle, you emerge emotionally stronger. Life has tested you and you were equal to the task. Then, when the next hurdle appears, you are better equipped to handle it. Problems and challenges bring out the best within you. You discover abilities you never knew you possessed. You would never have discovered these talents if life had not made you travel over some bumpy ground. Adversity reveals to you your own strengths and capacities, and beckons you to develop those qualities even further. Let me give an example. Since 1997 to 2000 I was managing a project for a Savings Bank in Spain. The project was an Infrastructure

Today is a Good Day!

project plus a Change project. I was the Project Manager and I had a team leader reporting to me from a consulting company to lead all the process changes, job changes and organizational communication. After three months the customer complained about the Change part and asked me to fire the team leader. I tried convincing my customer that it would be too risky, but finally because of political reasons the team leader from the consulting company was out of the project. I had to assume all the responsibilities. It was very difficult for me, but I was honest with my manager and with my customer. I learned a lot. I got the support from my customer on a daily basis and I achieved good results. My effort was worth it.

- **Adversity encourages you to make changes and take action.** Most people cling to old, familiar patterns regardless of how boring or painful their lives have become. It often takes a crisis or a series of difficulties to motivate them to make adjustments. Problems are often life's way of letting you know that you are off course and need to take corrective action.

- **Adversity teaches valuable lessons**. Take the example of a failed business venture: the entrepreneur may learn something that enables him or her to succeed spectacularly on the next venture.

- **Adversity opens a new door.** A relationship terminates and you go on to a more satisfying relationship. You lose your job and find a better one. In these instances, the "problem" is not a problem at all, but rather an opportunity in disguise. One door in your life has been shut, but there is a better one waiting to be opened.

- **Adversity builds confidence and self-esteem.** When you muster all of your courage and determination to overcome an obstacle, you feel competent and gain confidence. You have a greater feeling of self-worth and you carry these positive feelings into subsequent activities.

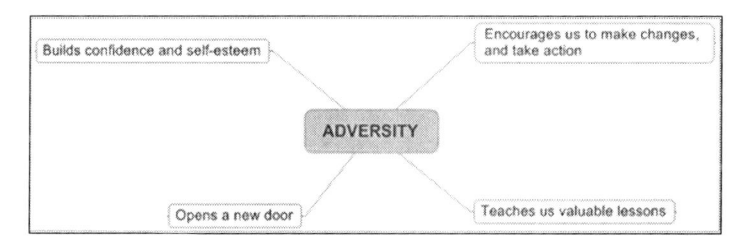

```
Builds confidence and self-esteem        Encourages us to make changes, and take action
                    ADVERSITY
Opens a new door                         Teaches us valuable lessons
```

Look for the Positive

Sure, you will have your share of problems and adversities in life. I am not suggesting that when tragedy strikes you deny your emotions or refuse to face reality. What I am saying is, do not immediately judge your situation as a tragedy and dwell on how bad off you are. Sometimes you will not be able to instantly spot the benefit that will come from being in your situation, but it does exist.

You always have a choice. You can view your problems as negative and become gloomy and depressed about them. Let me assure you this approach will only make things worse. Or, you can choose to see every seemingly negative experience in your life as an opportunity, as something you can learn from, as something you can grow from. Believe it or not, your problems are there to serve you, not to destroy you.

So, the next time you suffer a problem or setback in your life or in your project, don't get discouraged or give up. Do not let problems cloud your attitude window forever. Clear off that cloudy window. You may find, after the dust settles, that you can actually see better than you did before. Just remember the words of Napoleon Hill, "Every adversity carries with it the seed of an equivalent or greater benefit". For example: Two years ago I started a new personal project: to graduate in leadership at the Leadership Institute from PMI (Project Management Institute). I was admitted to the class. The training program had two possibilities to be followed: enroll in the US class or in the EMEA (Europe, Middle East and Africa) class. I participated in the first "face to face" session organized for professionals coming from EMEA. The second face to face session was scheduled for four months later, but in the mean time my mother suffered an accident (she broke her leg) and she had a surgery intervention. That accident happened just one day before the second face to face EMEA session was scheduled. Then I had to make a decision (go or not go). My decision was not to go and instead I stayed with my mother at the hospital until she recovered. Initially I was frustrated; a lot of effort, homework study and exercises, and reading appeared to be lost. I could not attend the second training session. The Training Program Director from the Leadership Institute called and suggested that I enroll in the second class for the US and attend their second "face to face" training session. So I did, and then I received the third training session, and completed the whole program during the rest of the year. I was graduated in Atlanta in October 2007. Because of that adversity, now I know double the number of professional colleagues worldwide. I had the huge opportunity to meet all colleagues from my first group in Europe and also to meet a second group of people in the US. It was great to increase my network and knowledge after a personal crisis.

Continually ask yourself what you have learned from your trying experience and focus on moving forward and growing as a person. In times of crisis, always strive to maintain an optimistic attitude and an open mind for this is the environment that will allow you to find the benefit in your difficulty.

Case Study

From 1997 to 2001, I was assigned as a Project Manager for a project in the South of Spain. I was living outside of home from Monday to Saturday every week during more than three years. The project was critical for the customer and we could not fail. This project generated a lot of stress, project and personal issues. I had a large project team—eight people from my company and more than thirty people subcontracted from different organizations.

After some months we achieved our first project milestone but afterwards people started to feel depressed and unmotivated. I analyzed the situation and understood that we were working more than twelve hours per day. We were not efficient. I led by example. I met the whole team and told them I would be leaving work at 6:30 p.m. every day, and I did. The result was really impressive; people followed me and acted in the same way. The project productivity increased dramatically.

On the other hand, I also found people on Monday morning telling me, *"Oh Alfonso. It's Monday, my God!, a long and sad week,…"*. I immediately told them, *"Don't worry, now "it's 9:30 am and in one hour we will do a coffee break. After the break lunch will be coming soon, and after lunch it's almost Tuesday…"*. I always tried to be positive with my people and transmit confidence and passion to get our job done. It was really helpful for us during those three years. Over the years I have discovered that professionals, develop special skills from difficult and stressing situations.

Summary

- Human beings and professionals appreciate the real important things in life when we live in difficult times.
- The road to success often travels through adversity.

Today is a Good Day!

- The project problems and issues you find when managing your projects are always opportunities to learn.

- Continually ask yourself what you have learned from your trying experience and focus on moving forward and growing as a person.

- In times of crisis, always strive to maintain an optimistic attitude and an open mind for this is the environment that will allow you to find the benefit in your difficulty.

CHAPTER 6

Your Words Make a Difference

Repeat anything often enough and it will start to become you.
—Tom Hopkins

When was the last time you seriously thought about the words you use each and every day? How carefully do you select them?

Now, you might be thinking, *"Alfonso, why all this fuss about words? What's the big deal?"* The answer is simple. Many times your words have much more power that what you can imagine. They can build a bright future, destroy opportunity or help maintain the status quo. Your words reinforce your beliefs, and your beliefs create your reality and contribute to project success.

Think of this process as a row of dominos that looks like this:

Thoughts – Words – Beliefs – Actions – Results.

Today is a Good Day!

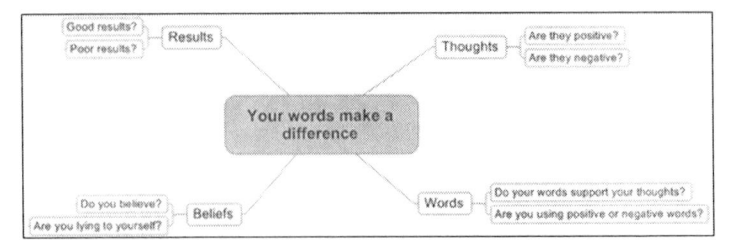

```
Good results?  ┐
               ├─ Results                    Thoughts ─┬─ Are they positive?
Poor results?  ┘                                       └─ Are they negative?

              ┌─────────────────────────┐
              │  Your words make a      │
              │      difference         │
              └─────────────────────────┘

Do you believe?    ┐                          ┌─ Do your words support your thoughts?
                   ├─ Beliefs        Words ───┤
Are you lying to   ┘                          └─ Are you using positive or negative words?
yourself?
```

Here is how it works. Henry has a ***thought***, such as *I am not very good when it comes to project sales*. Now, let's remember that he doesn't have this thought only once. Oh, no. He runs it through his mind on a regular basis, maybe hundreds or thousands of times in his life.

Then, Henry starts to use ***words*** that support this thought. He says to his friends and project management colleagues, *"I am never going to do very well in sales"* or *"I just hate making sales calls or approaching prospects"*. Here again, Henry repeats these phrases over and over … in his self talk and in his discussions with others.

This, in turn, strengthens his ***beliefs*** and it is at this stage where the rubber really meets the road. You see, everything that you will achieve in your life flows from your beliefs. So, in the sales example, Henry develops the belief that he's not going to be successful in project sales. This becomes embedded in his subconscious mind.

What can possibly flow from that belief? Because Henry does not believe in his project sales ability, he takes very little action, or he takes actions that are unproductive or ineffective. He doesn't do the things that would be necessary to succeed in project sales.

And then, quite predictably, Henry gets very poor ***results***. To make matters worse, Henry then starts to think more negative thoughts, repeat more negative words, reinforce negative beliefs and get even more negative results. It's a vicious cycle.

Of course, this whole process could have had a very happy ending if Henry had selected positive ***thoughts*** and reinforced them with ***positive words.*** In turn, he would reinforce the ***belief*** that he is successful in sales. As a result, Henry would take the ***actions*** consistent with that belief and wind up with outstanding ***results.***

Selecting the Right Words

Do not underestimate the role of your words in this process. Professionals who feed themselves a steady diet of negative words are destined to have a negative attitude. It is a simple matter of cause and effect. You can't keep repeating negative words and expect to be a high achiever. And that's because negative words will always lead to the reinforcement of negative beliefs and eventually to negative outcomes.

We usually repeat to ourselves things like, *"I am not good delivering presentations", "I am not good talking to upper managers", "I am not very good managing project cost".* And, after many years of using negative words, you develop a strong belief that you can't do these things.

Do you see how you create this situation by not being careful about the words you use? The truth is, you could eventually reverse this trend if you started using positive words about your ability to make repairs. This behavior depends on the motivational values from every professional. However, I have observed that many professionals are very negative by nature. If you belong to that group of professionals, now is the time to wake up and be conscious about the words you are using in your projects, with your people, with your customers, and with your managers. If you are a positive person, congratulations and welcome to the team, *"Today is a*

Good Day!" If you are not yet a member, you have the choice to change your attitude.

The State of Mind

In words we see the state of mind, character and disposition of the speaker. Years ago, my friend Randall L. Englund introduced me to Dr. Robert J. Graham, an experienced and worldwide recognized project professional, lovely, amazing, and friendly person. He is also a cultural anthropologist, and a university professor in the US, a consultant worldwide, and the author of four books on project management.

Graham has a physical challenge to contend with because he is in a wheel chair suffering from multiple sclerosis. However, over many years he has consistently delivered project management seminars, attended and been a lecturer and key note speaker in many Project Management Congresses, and consulted and assessed big multinational companies on the project management field. Graham loves people. He always speaks with enthusiasm; he transmits passion and security in all of his speeches. He can't use his arms very much, and he needs to rely on a power scooter or a wheel chair. He communicates through positive words, looking at you through his powerful smile. He has told me many stories, because he has travelled a lot worldwide. To this day he continues travelling coast to coast to visit his family and help with community affairs. It is incredible that someone in his physical state is not focused on himself. He is always trying to enjoy his life, never complaining.

By using positive and enthusiastic words, Robert Graham is empowering himself to achieve great things. He does not give any power to his limitations and, as a result, he is able to transcend them and accomplish more than many others.

Your Words Make a Difference

What obstacles are you facing in your life right now? Imagine the power you could unleash if you saw them as "just barely an inconvenience" instead of as an insurmountable barrier.

I have some suggestions to improve your positive words. To begin with, use positive self-talk as often as possible. In my opinion, the more the merrier. After all, you are talking to yourself, so you don't have to worry about others hearing your comments or judging what you say. The key is that you hear this positive input again and again, and it becomes deeply rooted in your subconscious mind.

Whether or not to share your goals with other people is a much trickier issue. One thing I have learned is this: *Never discuss your goals with negative people.* All they will do is argue and point out all the reasons why you will not be successful. Who needs that? Often, these *"negative nellies"* are the ones who do little or nothing in their own lives. They have no goals or dreams and they don't want anyone else to succeed either.

Yet, there are some instances when you can benefit by telling others about your goals. First of all, make sure that you are speaking with someone who is extremely positive and totally supportive of your efforts. This should be the kind of person who would be absolutely delighted if you achieved this goal and would do anything in his or her power to assist you. You may have a friend or colleague that fits this role or certain family members.

It is also important to share your goals with others who are working with you to achieve that outcome. For example, if a sales manager wants to increase sales in the coming year by 20%, he or she would make this goal known to everyone on the staff. Then, everyone can work together to achieve it.

Even though I am encouraging you to use positive words to move you toward your objectives, I am not suggesting that you ignore the obstacles that you may face or that you discourage feedback from other people. Before embarking on any goal, you want to prepare for what may be coming down the road. Personally I prefer to discuss those issues with someone who is positive, someone whose feedback includes creative solutions to the difficulties that may arise.

Furthermore, I'll only discuss my plans with people who are qualified to render an intelligent opinion on the subject. Let me give you an example. When I left HP to create my own company, I asked some colleagues their opinions about my decision. Most of them did not understand. Common opinions were *"too risky, you are fool, outside is raining too much"*. I only found one professional who encouraged me to do it, Luis M. supported my ideas, and gave me some contacts in the market to start up my business. He was a reviewer of my first project management book, and he is my friend.

I learned to listen to intelligent people, who are focused on the huge possibilities every human being has as a professional. A positive state of mind helped me a lot in my professional career.

Words and Accountability

There is another reason why, in some cases, you might decide to share your goals with someone else. In other words, if you tell others you are going to do something, then you have to go ahead and do it. Think of this approach as *"burning your bridges"*.

Let me assure you that I am not a believer in *"burning bridges"* when it comes to personal or business relationships. But sometimes the only way to move forward in life and to achieve an ambitious goal is to cut off all avenues of retreat.

This can be a very useful strategy. We may tell a friend that we are going to work out at the gym three times this

Your Words Make a Difference

week knowing that at the end of the week, this friend will ask whether we did. In fact, go to the gym three times.

Even more dramatic example is that of well-known motivational speaker Zig Ziglar. Ziglar decided to go on a diet and reduce his weight from 202 pounds to 165 pounds. At the same time, he was writing his book, *See You at the Top*.

In the book, Ziglar included a statement that he got his weight down to 165 pounds. This was ten months before the book went to press. And then he placed an order with the printer for 25,000 copies. Now remember, at the time he wrote these words, Ziglar actually weighed 202. He put his credibility on the line with 25,000 people.

By including a statement that he weighed 165 pounds, Ziglar knew that he had to lose 37 pounds before the book was printed. And he did. Use this strategy selectively. Limit it to those goals that are very important to you and where you are committed to go the distance. Is it risky? You bet it is. But it is a tremendous motivator.

Emotions

Our vocabulary affects our emotions, our beliefs, and our effectiveness in life. For example, let's say that someone has lied to you. You could react by saying that you are angry or upset. But if, you used the words, *"furious", "livid"* or *"enraged"*, your physiology and your subsequent behavior would be dramatically altered. Your blood pressure would rise. Your face would turn beet red and you would feel tense all over.

On the other hand, what if you characterized the situation as *"annoying"* or said that you were *"peeved"*? This lowers your emotional intensity considerably. In fact, saying that you are "peeved" will probably make you laugh and break the negative emotional cycle completely. You would be much more relaxed. Of course, you can also intentionally select

Today is a Good Day!

words to heighten positive emotions. Instead of saying, *"I am determined"*, why not say. *"I am unstoppable"*. Or, in place of declaring that you *"feel okay"*, try *"I feel phenomenal"* or *"I feel just tremendous"*.

Juicy, exciting words like that lift your spirits to a higher level and profoundly influence those around you. When you consciously decide to use such terms, you actually choose to change the path on which you are travelling. Others will respond to you differently, and you will alter your perception of yourself, as well.

Let's take a look at your life for a moment. Are there any areas where you have been using phrases like, *"I can't"*, *"I am not good at"*, or *"It is impossible?"* We all know project professionals who make statements like these:

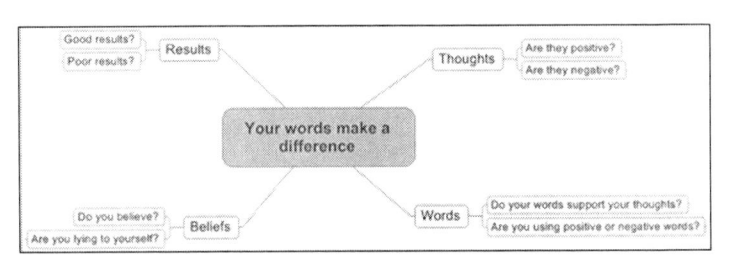

Diagram: "Your words make a difference" — Results (Good results? Poor results?), Thoughts (Are they positive? Are they negative?), Words (Do your words support your thoughts? Are you using positive or negative words?), Beliefs (Do you believe? Are you lying to yourself?)

When you make these comments day in and day out for ten to twenty years, you are programming your mind for failure. **It all comes back to your attitude.** Every one of these examples reflects a negative attitude. And if you see the world through a smudged window, you are going to use negative language and get disappointing results.

Fortunately, you can control your words, which means you have the ability to build a positive belief system to produce the results you want. The first step is awareness. Let's examine the phrases you have been using in four key areas of your project life – relationships, finances, career and health.

Your Words Make a Difference

1. **Relationships** - Do you say things like *"All good men (or women) are taken"* or *"People are always taking advantage of me?"* If you do, you are literally programming yourself for unhappy relationships. Your mind hears every word you speak and it sets out to prove you right. With regard to the above examples, your mind will see to it that you attract only those persons who will disappoint you or take advantage of you. Is this what you want? If not, stop repeating such negative statements.

2. **Finances** - What words do you use on a regular basis to describe your current financial situation and your prospects for the future? Phrases such as, *"I am always in debt"*, *"The economy is lousy"* or *"No one is buying"* work against you. It is far better to choose language which reaffirms prosperity and better economic times. Of course, you will not necessarily have abundant wealth within a few days after changing the way you speak. But the physical conditions can change only after your beliefs have altered. Clearing up your language is an important first step. After all, the people with wealth in this world did not get that way by thinking about being in poverty. And the people who always talk about a lack of money generally don't accumulate much of it.

3. **Career** - If I were to ask you about your career prospects over the next five to ten years, how would you respond? Be honest. Would you say that things will probably remain the same as they are now? Or would you describe a higher position with more challenges, more responsibilities and increased financial rewards? If you respond, *"I don't know where I am going in my career,"* chances are not much will change. Your language reflects your lack of vision and direction. If, on the

other hand, you have a clear goal which you can articulate fairly often, even if only to yourself, you stand an excellent chance of reaching that goal. The same, of course, holds true if you have your own business. Do you use language that is consistent with the growth of your business? Or do you constantly talk about how you will never get to the next level?

4. **Health** - Without question, our words have a profound impact on our health. For example, imagine that a group of us sat down to what seemed to be a perfectly wholesome and delicious meal. Then, two hours later I called and told you that every person who ate with us had been rushed to the hospital and treated for food poisoning. Suppose that you felt perfectly fine before I called. How would you react after hearing my information? Most likely, you would clutch your stomach, get pale and feel very ill. Why? Because my words instilled a belief in you which, in turn, your body started to act upon. This same bodily reaction would have occurred even if I was playing a cruel joke and was lying about the whole situation. Your body responds to words it hears you and other people say. That's why it makes absolutely no sense to keep repeating that you have *"chronic back pain that will never go away"* or that you get *"three or four bad colds every year"*. By uttering these statements, you actually instruct your body to manifest pain and disease. Please don't misunderstand. I am not suggesting that you deny pain or disease or that you can overcome any illness, but there is certainly nothing to be gained from using language that reinforces suffering and uncurability.

Up to you

Have you thought about the words you use in your personal and professional life? When we repeat certain phrases over and over, it is as if a "groove" is formed in our brain. We keep replaying the same old refrain in our heads like a broken record. The trouble is, whenever you say those words you just deepen the groove, replaying the same old myths in your mind, strengthening the same old beliefs, and getting the same old results.

Recognize, however, that just because you have said things in the past there is no reason to blindly continue doing so. While it takes some discipline and vigilance on your part to make changes in your language, it is well worth the effort. So, from now on, consciously choose words that will point you in the direction of your goals. Ask a friend to remind you when you slip.

Remember, it is up to you to speak in a way that will move you toward what you want in life and in the projects you are managing.

Nonverbal Communication

Nonverbal messages often contradict the verbal; often they express feelings more accurately than the spoken or written language. Numerous articles and books have been written on the importance of nonverbal messages. Some studies suggest that form 60% to 90% of a message's effect comes from nonverbal cues. I mean:

Appearance - Body language - Silence - Time and Space.

1. **Appearance** - It conveys nonverbal impressions that affect receivers' attitudes toward the verbal message even before they read or hear them. For instance, an envelope's appearance – size, color,

weight – may impress the receiver as *"important"*, *"routine"* or *"junk"* mail. Next, the letter, report, or title page communicates non-verbally before its contents are read by the kind of paper used, its length, format, and neatness. Finally, the language itself, aside from its content, communicates something about the sender. Is it carefully worded and generally correct in mechanics such as spelling grammar, and punctuation?

What is the effect on oral messages? Whether you are speaking to one person face to face or to a group in a meeting, personal appearance and the appearance of your surroundings convey nonverbal stimuli that affect attitudes toward your spoken words.

2. **Personal appearance** - Clothing, hairstyles, neatness, jewelry, cosmetics, posture, and stature are part of personal appearance. They convey impressions regarding occupation, age, nationality, social and economic level, job status, and good or poor judgment, depending on circumstances.

3. **Appearance of surroundings** - Aspects of surroundings include room size, location, furnishings, machines, architecture, wall decorations, floor, lighting, windows, and view. Surroundings will vary according to status and according to country and culture.

4. **Body language** - Included under body language are facial expressions, gestures, posture and movement, smell and touch, voice and sounds. The eyes and face are an especially helpful means of communicating non-verbally. They can reveal hidden emotions, including anger, confusion,

enthusiasm, fear, joy, surprise, uncertainty, and others. They can also contradict verbal statements. For example: because he was embarrassed, a new team member answered "yes" when asked by his Project Manager if he understood his instructions. Yet, the Project Manager should have noticed the employee's bewildered expression and hesitant voice and restated the instructions more clearly.

In the United States, direct eye contact is, (but not staring), is considered desirable when two people talk. The person whose eyes droop or shift away from the listener is thought to be either shy or dishonest and untrustworthy. However we must keep in mind that it's not seen exactly the same way in other cultures, and it also depends on the situation.

In some occupations, actions speak louder than words. Gestures and movements are culture-specific. The meaning of a gesture in the United States may be completely different in Europe and Asia. In the United States, a clenched fist pounding on the table can indicate either anger or emphasis. Such a display in Asia would be unacceptable. Continual gestures and movement such as pacing back and forth may signal nervousness and may be distracting to listeners. Handshakes reveal attitudes or sometimes handicaps, by their firmness or limpness.

Legs also communicate nonverbal messages. Consider, for example, a man sitting with his legs stretched out on top of his desk during an interview; a person shifting from one leg to the other in rhythmic motion; a woman pacing back and forth while speaking. Posture and movement can convey self-confidence, status or interest. A confident executive may have a relaxed posture, and yet may stand more erect than a timid subordinate. An interested listener may lean forward toward the speaker while one who is bored may lean away, slump, or glance at the clock.

Various scents and fragrances sometimes reveal the emotions of the sender and sometimes affect the reactions of the receiver, especially if the receiver is sensitive to scents. Touching people can also communicate friendship, love, approval, hatred, anger, or other feelings.

Your voice quality and extra sounds you make while speaking are also a part of nonverbal communication called paralanguage. Paralanguage includes voice volume, rate, articulation, pitch, and the other sounds you may make, such as throat clearing and sighing. The words, "You did a great job on this project" could be a compliment. But if the tone of voice is sarcastic and said in the context of criticism, the true meaning is anger.

A loud voice often communicates urgency while a soft one is sometimes calming. Speaking fast may suggest nervousness or haste. A lazy articulation, slurring sounds or skipping over syllables or words, may reduce credibility. A lack of pitch variation becomes a monotone, while too much variation can sound artificial or overly dramatic. Throat clearing can distract from the spoken words. Emphasizing certain words in a sentence can purposely indicate your feelings about what is important.

Silence, time and space can communicate more than you may think, even causing hard feelings, loss of business, and profits. It pays to know these differences across cultures. Suppose you wrote a request to your supervisor for additional funds for a project you were developing. If you receive no answer for several weeks, what is your reaction? Do you assume that the answer is negative? Do you wonder if your supervisor is merely very busy at the moment and has been unable to answer your request? Do you think your supervisor is rude or considers your request unworthy of an answer?

Concepts of time, however, vary across cultures and even in the United States. Americans and Germans, for example, are quite punctual. Mid Eastern business people think little of arriving after an agreed-upon time, not out of discourtesy, but rather a feeling that the task will be accomplished regardless of time. If you arrived on time for a meeting in Spain, your host might wonder why you came so early.

Case Study

Some years ago I managed a project for the Spanish Foreign Ministry. It was an infrastructure project focused on creating the right hardware and software architecture to encode and decode messages between worldwide Embassies. The IT manager (José) had very clear requirements for that project, but he never wrote them down. We did it for him. I had the perception that he was a difficult person, very rude, and prone to speaking and acting out a negative behavior.

I still remember when he said that I was always smiling, and he was disappointed about that. However, his people worked well with me and my team. We took care of using respectful sentences working with them, and they appreciated the difference between us and their boss.

One day we had a database problem and we needed to work overnight with the whole team. The customer IT manager was absent, but when he returned the next morning, his first sentence was, *"Hello, are you working or talking as always?"* Can you imagine how everyone who heard that sentence felt?

The database problem was corrected and the project, although with some time delay, finished according to cost and requirements. The IT customer was happy about the results and we had lunch together. At the end of the lunch the IT manager said to the team, *"Congratulations for the project success, but you delivered the project overtime"*.

I continued working for that customer with my team on other projects. After some years I gained credibility with this customer and from time to time, we went to lunch together. At one of the lunches he told me, *"Alfonso, my people respect you very much. They are happy when you come to my office. What do you do?"* I answered him, *"I listen to them, I respect them, and I tell them that I need them for achieving project success. I smile frequently because every day is a good day"*. I told him, *"José, spend more time with your people, talk to them, don't be negative with them, be more aware of the words you use with them and appreciate their efforts and achievements personally. The return of your investment will be significant"*.

I feel happy about this story because now this customer meets me with a big smile every time I see him. He has changed his behavior to be more positive and his people appreciate him so much more for it.

Summary

- Remember that many times your words have much more power than what you can imagine. They can build a bright future, destroy opportunity or help maintain the status quo. Your words reinforce your beliefs, and your beliefs create your reality, and then contribute to your project.

- You can't keep repeating negative words and expect to be a high achiever. And that is because negative words will always lead to the reinforcement of negative beliefs and eventually to negative outcomes.

Your Words Make a Difference

- Our vocabulary affects our emotions, our beliefs, and our effectiveness in life. Fortunately, you can control your words, which means you have the ability to build a positive belief system, and to produce the results you want. The first step is awareness.

- Remember, it is up to you to speak in a way that will move you toward what you want in life and in the projects you are managing.

Today is a Good Day!

CHAPTER 7

How Are You?

> *Your day goes the way the corners of your mouth turn*
> —Unknown

Our answer to the question *"How are you?"* may seem like a small thing. But we must answer that question many times every day. Then, it's not a small thing at all. It is a significant part of our daily conversations.

When someone asks, *"How are you?"*, What do you say? Your answer is usually only a few words. And yet, that short response tells a lot about you and your attitude. In fact, your response can literally shape your attitude.

I have observed that the responses to, *"How are you?"*, can be classified into three categories: **negative, mediocre, and positive**. Let's examine these three categories and some common responses under each one.

1. **Negative answers** - The negative reply to, *"How are you?"* may include phrases like:
    - "Don't ask"
    - "Lousy"
    - "Monday again"
    - "Terrible"

Today is a Good Day!

- "It's not a good day"
- "Bad, as always"
- "I'm tired"

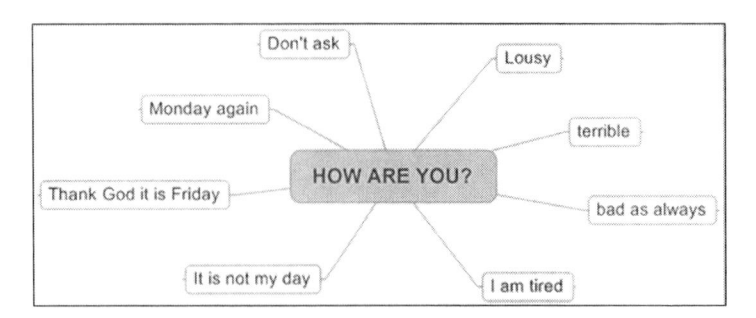

```
                Don't ask              Lousy
        Monday again
                                                 terrible
                          HOW ARE YOU?
Thank God it is Friday
                                            bad as always
                It is not my day       I am tired
```

When a Project Manager professional answers with, *"Don't ask"*, I know I am in for trouble. That person is going to unleash a multitude of complaints and make me sorry for asking the question in the first place.

And I really pity those who take the, *"Thank God it's Friday"* approach to life. Think of what they are saying. *"Monday, Tuesday, Wednesday and Thursday"* are bad days every week. For these people, four fifths of their work week is lousy. The fifth day, Friday, is *"bearable"* only because they know they will have the next two days off. Is this a way to live your projects and your life? Are you beginning to see how these negative phrases can poison your attitude and turn off your project team and other project stakeholders?

2. **Poor answers** - Those in the mediocre group are a step up from the negative bunch but they still have plenty of room for improvement. Here are some of the things they say:

    - "I am okay"
    - "Not too bad"
    - "It could be worse"

How Are You?

- "Every day older and older"
- "I am fine"

Do you really want to spend a lot of time with someone who thinks that life is *"not too bad"*? Is that the person you want to do business with? When we use words like these, we also diminish our energy. Can you imagine someone saying, *"could be worse"* with an upright posture and a lot of enthusiasm? Of course not. These people sound like they have not slept in two days.

There is no getting around it. People who use mediocre words will develop a mediocre attitude, and get mediocre project results. And I know you don't want that.

3. **Positive answers** - Passionate people say enthusiastic responses, like:

- "If I was better I would have a twin"
- "I am terrific!"
- "Superb!"
- "I am fantastic!"
- "Excellent!"
- "Great!"
- "Today is a Good Day!"

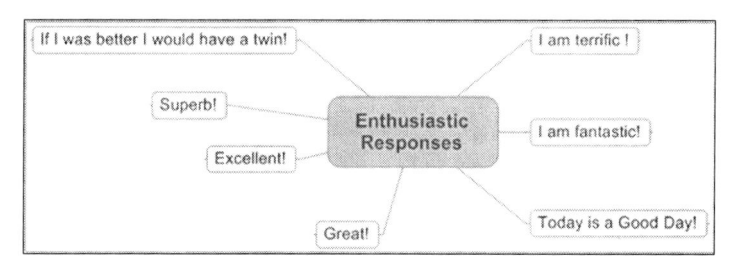

Enthusiastic Responses
- If I was better I would have a twin!
- I am terrific!
- Superb!
- I am fantastic!
- Excellent!
- Great!
- Today is a Good Day!

Those who use positive words like these have a bounce in their step. You feel a little better just by being around them. Be honest. How did you feel as you read the positive list? I don't know about you, but I am energized and excited as I review that list. These are the people I look forward to meeting today. These are the people who are more likely to get my business. Why not go back and re-read the negative list and the mediocre list? Say them out loud. How do they make you feel? Bummed out, for sure.

You see, if given the choice, I would rather be around people who are positive and full of life as opposed to those who are negative and listless. It is like the old saying that everybody lights up a room – some when they walk into the room, and some when they walk out. You want to be the one who lights up a room when you walk in.

As for me, when someone asks me, *"How are you?"* I usually respond with, *"Very good, today is a good day"*. It projects a positive attitude to the other person and the more I say it, the more I feel better.

Join the Positive Professionals

Well, you have had a chance to review some typical response in each category – negative, mediocre and positive. Which of these phrases do you use most often? Which responses do your friends and family use?

If you find yourself in the negative or mediocre group, I suggest you immediately consider revising your response and joining the ranks of the positive. Here is why. When you are asked how you are and you say horrible or not too bad, your physiology is adversely affected. You tend to slump your shoulders and head and take on a depressed posture.

What about your emotions? After stating that you are lousy, do you feel better? Of course not. You feel even more down in the dumps because negative words and thoughts generate negative feelings, and eventually, negative results. It is up to you to break it. Even if real circumstances in your life persuaded you to state that you are lousy – perhaps a promising business deal fell through, or your child received poor grades in school – your gloomy attitude does nothing to improve the situation. To make matters worse, your mediocre or negative reply turns others off; they are dragged down just being around you and hearing your pessimism.

Practice a New Approach

If all of these negative consequences flow from your words, why do you continue to say them? More than likely, it is because you have not recognized that you have a choice in the matter. Instead, you are following a habit that you developed many years ago, a habit that no longer serves you.

In the end, your own words are a self-fulfilling prophesy. If you say, *"Everything is terrible"*, your mind is attracted to those people and circumstances that will cause that statement to be true. If, on the other hand, you repeatedly state that your life is wonderful, your mind will begin to move you in a positive direction.

For instance, just consider what happens when you respond that you are, *"Excellent or Terrific"*. As you say these words, your physiology begins to correspond with your optimistic language. Your posture is more upright. Other people are attracted to your energy and vitality. Your business and personal relationships improve. Will all of life's problems magically disappear? No, but you have set in motion a very important principle. We get what we expect in life.

I can tell you from firsthand experience that this is one of those little things in life that makes a big difference. About fifteen years ago, when someone asked me, *"How are you?"*, I

Today is a Good Day!

said something like, *"Okay"* with very little energy. What was I doing? I was programming myself to have *"okay"* relationships with people. I was programming myself to have *"okay"* success. I was programming myself to have an *"okay"* attitude and *"okay"* life.

But then, through an experiment, I learned that I didn't have to settle for an *"okay"* life. I picked my response up a few notches and began to say, *"Terrific"*. I said it with energy. Sure, at first it was a little uncomfortable. Some people looked at me like I was a little strange. But after about a week, it started to come naturally. I was amazed at how much better I felt and how people were much more interested in talking with me.

Believe me, this is not rocket science. You don't need talent, money or good looks to have a great attitude. You just need to get in the habit of using a high-energy, positive response, and you will get the same exciting results I got.

What Happens if I Feel That Today is Not a Good Day?

Whenever I do a presentation or a Seminar I say, *"Today is a good day"*. Many people are surprised at the beginning. Sometimes, at the end of the event some people came to me and ask, *"What if today is not a good day?"*.

I don't want to lie to my customers and colleagues by telling them everything is wonderful when it's not. Now don't get me wrong. I put the highest value on integrity and telling the truth. Yet, I don't think this is a matter of telling the truth. Let me explain.

Assume for a moment that Rose feels tired. When someone at work asks her, *"How are you?"*, she wants to be perfectly honest, so she says, *"I am tired"*. Here is what will happen. Rose will reinforce the belief that she is tired. She will feel even more fatigued. She will probably slump her shoulders and let out a sigh. She will have a lousy, unproductive day at work.

Let's get back to the person who asked Rose the question – and who probably regrets it now. That person also feels worse. After all, when someone tells you how tired she is, do you feel uplifted? No way. Just the suggestion of the world "tired" and you start yawning. So, Rose has brought herself down, as well as her co-worker.

Okay, Rose goes home after her gruelling day and now she is exhausted. So she plops into her favorite chair and opens the newspaper to look at the winning lottery numbers. As she pulls her ticket out of her wallet, she discovers that she is holding the winning ticket. She just won $10 million.

What do you think Rose would do? Remember, she was very tired. You and I both know that Sally would leap out of her chair, jump up and down, while screaming and waving her arms in the air. You would think she was leading an aerobics class. Naturally, she would run to pick up the phone to call her family and friends. She would be a bundle of energy and would probably stay up all night celebrating and planning what to do with the money.

But wait a minute. Ten seconds ago Rose was exhausted. Now, she has the energy of a fifteen-year-old cheerleader who has just been told she made the cheer leading squad. What happened in those ten seconds to change her from feeling utterly exhausted to wildly exuberant? Did she get a shot of vitamin B-12? Did anyone throw a bucket of ice water in her face?

No. *Her transformation was entirely mental!*

Now I am not trying to diminish what Rose was feeling. Her fatigue was very real, but it was not as much physical as mental. So, was Rose telling the truth when she said she was tired? It really had very little to do with the truth. It was a matter of what Rose wanted to focus on. She could concentrate on feeling tired. That was one option. On the other hand; she could have thought about the many blessings in her life and felt very fortunate and energized.

How we feel is very often a subjective matter. When we tell ourselves that we are tired, we feel tired. When we tell ourselves that *"today is a good day"*, we feel energized. We become what we think about.

Use Your Passion to Answer

Try this experiment. When anyone asks, *"How are you?"*, try to respond with energy and enthusiasm. Say that you are, *"Great"* or *"Today is a good day!"* Say it with a smile and a sparkle in your eye. It does not matter whether or not you completely and totally feel terrific at that moment. Simply apply the act-as-if principle. In other words, if you want to be more positive, act-as-if you already are and, pretty soon you will find that you have, in fact, become more positive.

Do not worry if you feel a little uncomfortable saying these words at the beginning. Stick with it and eventually you will grow into it. You will quickly notice that you feel better, that others want to be around you, and that positive results will come your way. Then, *"How are you? Today is a good day"*.

Case Study

At the age of thirty three, I was a Project Manager working in a project far from my city of residence. When a manager asked me about the project I managed, I always answered, *"Oh my God! I am so frustrated because of the lack of support of my organization for this particular project. I also feel frustrated with my customer. He is not collaborating....."*

All those answers are negative, but I wasn't conscious of that. I wasn't aware that my team was listening to my answers too. They observed my negative attitude. I generated a lot of frustration and lack of motivation in my team and I didn't know it.

I was very lucky. One weekend I met a manager from another organization in a restaurant and had the opportunity to drink a cup of coffee with him. I told him about my project situation. He gave me great ideas and advice to move forward and to improve my attitude. He said, "*Alfonso, there is no project without issues or problems. It is like life, the point is that you need to deal with those problems and don't forget that you work for an organization; you must ask them for their support. You must communicate to all project stakeholders and tell them how much you need them for achieving project success. I am sure you will have down moral moments, but please be focused on the great things you have. Thinking about your family is OK, think about you can do it, and be positive. When somebody is asking you about your project say, "It is controlled. I have some issues to be managed but if I need help, I'll let you know. That answer is more positive, it inspires optimism, energy and enthusiasm. Many times there is a lack of energy in some projects.*"

That answer empowered me a lot. It was not easy to use that advice immediately, but I did it step by step. Now I apply my approach every day, because *"Every day is a Good Day!"*

For sure you will be able to choose between the positive part of your project or the negative one. I suggest you choose the *"positive"*.

Summary

- When someone asks, *"How are you?"*, What do you say? Your answer is usually only a few words. And yet, that short response tells a lot about you and your attitude. In fact, your response can literally shape your attitude

- As for me, when someone asks me, *"How are you?"*, I usually respond with *"Very good, today is a good day"*. It projects a positive attitude to the other person and the more I say it, the more I feel better.

- If you find yourself in the negative or mediocre group, I suggest you immediately consider revising your response and joining the ranks of the positive.

- You don't need talent, money or good looks to have a great attitude. You just need to get in the habit of using a high-energy, positive response, and you will get exciting results.

- How we feel is very often a subjective matter. When we tell ourselves that we are tired, we feel tired. When we tell ourselves that we feel terrific, we feel energized. We become what we think about.

- Do not worry if you feel a little uncomfortable saying these words at the beginning. Stick with it and eventually you will grow into it. You will quickly notice that you feel better, that others want to be around you, and that positive results will come your way. Then, *"How are you?"* becomes *"Today is a good day!"*

CHAPTER 8

Don't Complain About Your Projects

> *If you have time to whine and complain about something then you also have the time to do something about it.*
> —Anthony J. D'Angelo

My father said that a pessimistic person is an optimistic one well informed. How do you feel when someone unloads all of his problems and complaints on you? Not very happy and energized, is it? The truth is, nobody likes to be around a complainer – except, perhaps, other complainers.

Of course, all of us complain at one time or another. The important question is: How often do you complain? If you are wondering whether you complain too much, simply ask your colleagues. They will let you know.

Now, when I say, *"complain"*, I am not talking about those instances when you discuss your problems in an attempt to search for solutions. That's constructive and commendable. And I am not referring to those occasions when you share your project life experiences with colleagues or friends in the context of bringing them up to date on the latest developments in your professional life. After all, part of being human is sharing our experiences and supporting each other.

Nobody Wants to Hear About Your Pains

One of the most common areas of complaining is the subject of illness. In this category are comments such as, *"My back is killing me"* or *"I have a terrible headache"*. Worse yet, some people get very graphic in explaining the gory details of their particular ailment. What can I do for you if you have a stomach ache? I am not a physician; you must go to the doctor if you have a medical problem. More importantly, why are you telling me this? You might want sympathy, but all you are doing is dragging me down and reinforcing your own suffering. Talking about pain and discomfort will only bring you more of the same and encourage those around you to look for the exits.

When it comes to complaints about projects, the principle of escalation usually rears its ugly head. Here is how it works. You tell your colleague about the problems you've had in your project. Your colleague interrupts you and speaks of, the agony you went through with the project. Your friend interrupts and says, *"You think you had it bad. When I managed my last project, the first day, I had a 40 Celsius degree fever and had to be rushed to the hospital. I almost died"*. Or, tell someone that your back or foot hurts – and count how many seconds it takes for that person to switch the conversation to his or her own back pain and aching feet. Complainers love to play this game. Their pain is always worse than yours.

Some People Had Some Reasons to Complain

I was in my office and thinking about some of the activities of my project that were not going as well as I had planned. You know the typical project problems – results not happening as fast as I had expected. And I'll confess that I had been doing a little complaining about it.

Then José walked in. (José is in his early 30's and came to Spain about six years ago from Nicaragua. He works for a company that cleans homes and offices.) You talk about positive attitude; José is one of the most positive people I have ever met. He is always smiling and upbeat.

On this day however, I asked José about the last earthquake and its impact on his homeland. The smile quickly left his face. He told me of the devastation the earthquake had caused. Thousands of people had died. José said that his father, mother and brother still lived in Nicaragua, and he had no idea if they were dead or alive. He had no way to contact them. All the phone lines in that area had been destroyed. José said he thought about his family every day.

Can you imagine the agony of not even knowing if your family is still alive?

Then José went on to tell me about all the things he was doing to help the people in Nicaragua. He was collecting money, clothing and other necessities. He was actively working with relief organizations. Instead of just griping about the problem, he was doing whatever he could to ease their pain.

After speaking with José, I began to realize just how inconsequential my own problems were and how fortunate I was. You better believe I stopped complaining. I faced the rest of the day with renewed energy and a much better attitude.

By the way, several weeks later I saw José again. And yes, he had his usual winning smile and his great attitude. The good news is that his family members are all alive. The bad news is that they lost everything in the earthquake. I can not even fathom what it is like to lose everything you own and have to start over from scratch, especially under those difficult conditions. José has every reason to complain about his family's bad luck. But he doesn't. He realizes that complaining would be a terrible waste of his time and energy. Thank you José, for reminding us all that complaining is not the answer to our challenges in life.

Putting Things in Perspective

There is another valuable lesson that we can learn from José, and that is the importance of keeping things in perspective. Over the years, I have noticed that complainers lack perspective, they tend to blow their problems way out of proportion.

Optimistic people tend to have a sense of what is truly important in life. Think about the people you know. Do you have any friends who get bent out of shape because they got a flat tire?

It is clear these folks have lost sight of the "relative importance" of things. I think we can all learn from Eddie Rickenbacker, who drifted in a life raft for 21 days, hopelessly lost in the Pacific. After surviving the ordeal, Rickenbacker said, "If you have all the fresh water you want to drink and all the food you want to eat, you ought never to complain about anything.

Let me share with you some of the things I am grateful for:

- I am in good health.
- My family is in good health.
- I have my own home.
- We have some food to eat and clean water to drink.
- I live in Spain and enjoy our country.

- I love my work and my profession.
- I love to travel and meet new and fascinating people.
- I have some loyal friends.

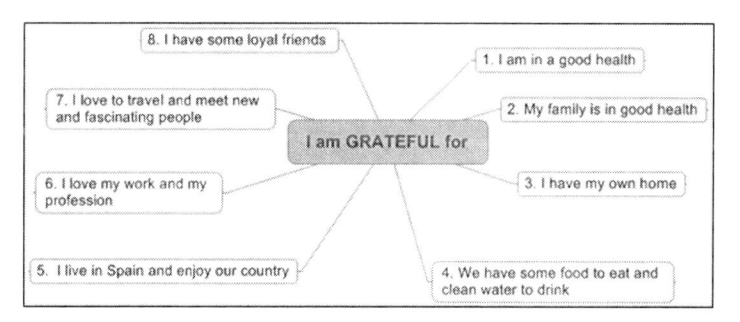

```
I am GRATEFUL for:
1. I am in a good health
2. My family is in good health
3. I have my own home
4. We have some food to eat and clean water to drink
5. I live in Spain and enjoy our country
6. I love my work and my profession
7. I love to travel and meet new and fascinating people
8. I have some loyal friends
```

This is just a partial list of the blessings in my life. Even with all of these wonderful things, there are times when I start to take some of them for granted. But I have learned to quickly re-connect with these blessings and it boosts my attitude and brings me back on course.

So, what is it that you have been complaining about lately? Are they really *"life and death" matters?* The next time you are tempted to gripe about your problems, pick up a pen and piece of paper and start listing all the reasons you have to be grateful. Then, please put all your project issues and problems in perspective.

The Positive News Generator

I am not suggesting that you just sit back and ignore all of the problems in your life or in your projects. However, rather than complaining, it is far better to focus your attention and your energy on those steps you can take to solve, or at least lessen, your problem. For instance, let's say you are feeling a little tired

Today is a Good Day!

lately. Instead of telling everyone how lousy you feel, make an effort to exercise more regularly or get to bed a little earlier.

To review: Complaints works against you in three ways.

1. No one wants to hear negative news about your illness and your problems.
2. Complaining reinforces your own pain and discomfort. So why keep replaying painful, negative memories?
3. Complaining, by itself, accomplishes nothing and diverts you from the constructive actions you could be taking to improve your situation.

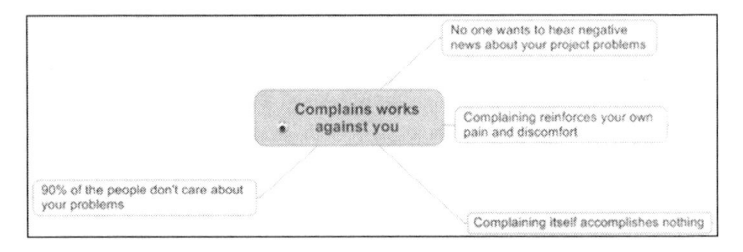

Diagram: Complains works against you
- No one wants to hear negative news about your project problems
- Complaining reinforces your own pain and discomfort
- 90% of the people don't care about your problems
- Complaining itself accomplishes nothing

It is been said that 90% of the people don't care about your problems, and the other 10% are glad you have them. Seriously though, all of us can cut down on our complaining. From now on, let's do ourselves and others a favor and make our conversations uplifting.

The people who don't complain very much (and those who speak positively) are a joy to be around. Decide to joint that group – so people won't have to cross the street when they see you coming. Perhaps the most important step in stopping the habit of complaining is to disconnect the undesirable behavior from your identity. A common mistake the chronic complainer makes is to self-identify with the negative thoughts running through their minds. Such a person might admit, *"I* know I'm responsible for my thoughts, but I don't know how to stop myself from thinking negatively so often". That seems like a step in the right direction, and to a certain degree it is, but it's also a trap. It's good to take responsibility for your thoughts, but you don't want to identify with those thoughts to the point where you end up blaming yourself and feeling even worse.

A better statement might be, *"I recognize these negative thoughts going through my mind. But those thoughts are not me" As I raise my awareness, I can replace those thoughts with positive alternatives".* You have the power to recondition your thoughts, but the trick is to keep your consciousness out of the quagmire of blame. Realize that while these thoughts are flowing through your mind, they are not you. You are the conscious conduit through which they flow.

Mental Conditioning

Although your thoughts are not you, if you repeat the same thoughts over and over again, they will condition your mind to a large extent. It's not completely accurate to say that we become our dominant thoughts, I think that's taking it a bit too far.

Consider how the foods you eat condition your body. You aren't really going to become the next meal you eat, but that meal is going to influence your physiology, and if you keep eating the same meals over and over, they'll have a major impact on your body over time. Your body will crave and expect those same foods. However, your body remains separate and distinct from the foods you eat, and you're still free to change what you eat, which will gradually recondition your physiology in accordance with the new inputs.

This is why negative thinking is so addictive. If you keep holding negative thoughts, you condition your mind to expect and even crave those continued inputs. Your neurons will even learn to predict the reoccurrence of negative stimuli. You'll practically become a negativity magnet.

    Let's imagine you must present a new project proposal for a customer, and from the beginning you think about the customer not liking your proposal repeatedly. During your proposal presentation you will get more and more nervous, so you will not be able to sell your proposal added value correctly to your customer, and you probably will not be able to win that project proposal. So be positive.

The Trap of Negative Thinking

This is a tough situation to escape because it's self-perpetuating, as anyone stuck in negative thinking knows all too well. Your negative experiences feed your negative expectations, which in turn attract new negative experiences.

    In truth, most people who enter this pattern never escape it in their lifetime. It's just that difficult to escape. Even as they rail against their own negativity, they unknowingly perpetuate it by continuing to identify with it. If you beat yourself up for being too negative, you're simply reinforcing the pattern, instead of breaking out of it.

    I think most people who are stuck in this trap will remain stuck until they experience an elevation in their consciousness. They have to recognize that they're trapped and that continuing to fight their own negativity while still identifying with it, is a battle that can never be won. Think about it. If beating yourself up for being too whiny was going to work, wouldn't it have worked a long time ago? Are you any closer to a solution for all the effort you've invested in this plan of attack?

    Consequently, the solution I like best is to stop fighting and surrender. Instead of resisting the negativity head-on, acknowledge and accept its presence. This will actually have the effect of raising your consciousness.

Overcoming Negativity

You can actually learn to embrace the negative thoughts running through your head and thereby transcend them. Allow them to be, but don't identify with them because those thoughts are not you. Begin to interact with them like an observer.

It's been said that the mind is like a hyperactive monkey. The more you fight with the monkey, the more hyper it becomes. So instead, just relax and observe the monkey until it wears itself out of consciousness. Also recognize that this is the very reason you're here, living out your current life as a human being. Your reason for being here is to develop your consciousness. If you're mired in negativity, your job is to develop your consciousness to the point where you can learn to stay focused on what you want, to create positively instead of destructively. It may take you more than a lifetime to accomplish that, and that's OK. Your life is always reflecting back to you the contents of your consciousness. If you don't like what you're experiencing, that's because your skill at conscious creation remains underdeveloped. That's not a problem though, because you're here to develop it. You're experiencing exactly what you're supposed to be experiencing so you can learn.

Conscious Creation

If you need a few more lifetimes to work through your negativity, you're free to take your time. Conscious creation is a big responsibility, and maybe you don't feel ready for it yet. So until then you're going to perpetuate the pattern of negative thinking to keep yourself away from that realization. You must admit that the idea of being the primary creator of everything in your current reality is a bit daunting. What are you going to make of your life? What if you screw up? What if you make a big mess of everything? What if you try your best and fail? Those self-doubts will keep you in a pattern of negativity as a way of avoiding that responsibility.

Unfortunately, this escapism has consequences. The only way true creators can deny responsibility for their creations is to buy into the illusion that they aren't really creating any of it. This means you have to turn your own creative energy against yourself. You're like a god using his powers to become powerless. You use your strength to make yourself weak.

The reason you may be stuck in a negative thought pattern right now is that at some point, you chose it. You figured the alternative of accepting full responsibility for everything in your reality would be worse. It's too much to handle. So you turned your own thoughts against yourself to avoid that awesome responsibility. And you'll continue to remain in a negative manifestation pattern until you're ready to start accepting some of that responsibility back onto your plate.

Negativity need not be a permanent condition. You still have the freedom to choose otherwise. In practice this realization normally happens in layers of unfolding awareness. You begin to accept and embrace more and more responsibility for your life.

Assuming Total Responsibility

You see … the real solution to complaining is accepting responsibility. You must say to the universe (and mean it), *"I want to accept more responsibility for everything in my experience."*

Here are some examples of what I mean by accepting responsibility:

Don't Complain About Your Projects

- If I am unhappy it's because I am creating it.
- If there is a problem in the world that bothers me, I'm responsible for facing it.
- If someone is in need, I am responsible for helping them.
- If I don't like my present circumstances, I must end them.
- If I want certain people in my project teams and initiatives, I must attract and invite them to be with me.
- If I want something, it is up to me to achieve it.

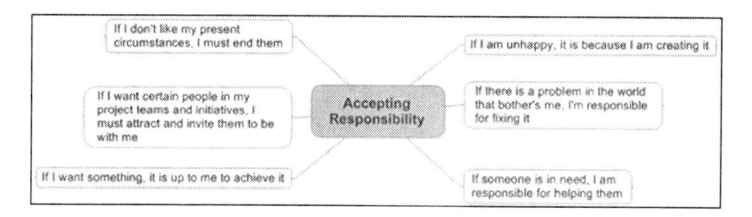

![Diagram: Accepting Responsibility with branches: "If I don't like my present circumstances, I must end them"; "If I am unhappy, it is because I am creating it"; "If I want certain people in my project teams and initiatives, I must attract and invite them to be with me"; "If there is a problem in the world that bother's me, I'm responsible for fixing it"; "If I want something, it is up to me to achieve it"; "If someone is in need, I am responsible for helping them"]

On the flip side, it may also help to take responsibility for all the good in your life. The good stuff didn't just happen to you. You created it. Well done.

Pat yourself on the back for what you like, but don't feel you must pretend to enjoy what you clearly don't like. But do accept responsibility for all of it ... to the extent you're ready to do so.

Complaining is the denial of responsibility. And blame is just another way of excusing yourself from being responsible. But this denial still wields its own creative power.

Conscious creation is indeed an awesome responsibility. But in my opinion it's the best part of being human. There's just no substitute for creating a life of joy, even if it requires taking responsibility for all the unwanted junk you've manifested up to this point.

When you catch yourself complaining, stop and ask yourself if you want to continue to deny responsibility for your reality or to allow a bit more responsibility back onto your plate. Maybe you're ready to assume more responsibility, and maybe you aren't, but do your best to make that decision consciously. Do you want sympathy for creating what you don't want, or do you want congratulations for creating what you do want?

Case Study

Eric was assigned as a Project Manager for a Software Migration project. Eric was a senior Project Manager who had spent fifteen years managing projects in the IT industry. However, Eric was always complaining about his project pains.

Unfortunately, Eric couldn't participate at the project requirements analysis, so he had to deal with the solution design that was generated by two consultants from the same company. Eric started the project delivery complaining everywhere about the lack of requirements definition, about the lack of documentation, and the lack of Sponsorship. But Eric never did anything to change the project situation. He never tried to talk to the manager about wanting or needing support. He never tried to talk to the colleagues who did the requirement analysis by asking for more data. He complained continuously without taking action. The project was delayed and he still complained, while his team became more and more frustrated. Then the customer started to worry about the chaotic situation. However, he was fortunate because during a weekend he met other Project Manager colleague and he talked to him about his project situation. His colleague suggested Eric change his behavior, explaining to him that he was not alone in their organization, that he could ask for help from the manager and that he needed to demonstrate to his customer and organization that he would be able to achieve the project success.

Some weeks passed and Eric met the manager. He explained how he needed help and support. The manager immediately promised to talk to the customer and have a meeting with the consultants that gathered the project requirements. Things started to change, and project status changed little by little. That was a positive lesson learned by Eric, who shared it with other Project Manager colleagues from his organization.

Summary

In summary I want to recap some of the ideas I have told you about in this chapter:

- The important question is: How often do you complain? If you are wondering whether you complain too much, simply ask your colleagues. They will let you know.

- Over the years, I have noticed that complainers lack perspective, they tend to blow their problems way out proportion.

- Optimistic people tend to have a sense of what is truly important in life. Think about the people you know.

- The people who don't complain very much (and those who speak positively) are a joy to be around. Decide to joint that group – so people won't have to cross the street when they see you coming

- Although your thoughts are not you, if you repeat the same thoughts over and over again, they will condition your mind to a large extent.

Today is a Good Day!

- It's been said that the mind is like a hyperactive monkey. The more you fight with the monkey, the more hyper it becomes. So instead, just relax and observe the monkey until it wears itself out of consciousness.

- When you catch yourself complaining, stop and ask yourself if you want to continue to deny responsibility for your reality or to allow a bit more responsibility back onto your plate.

CHAPTER 9

Associate With Positive Professionals

You have to perform at a consistently higher level than others. That's the mark of a true professional.

—Joe Paterno

In High School, Emile spent a lot of time with a bunch of guys in his neighbourhood. According to Emile, these guys just liked to sit on the front porch and watch cars go by. They had no goals and no dreams and they were always negative.

Whenever Emile suggested they do something new, the others would discourage him. *"It is stupid"* or *"not cool"*, they told him. Mike just went along with them so he could remain part of the group.

When Mike went off to college, he still ran into some negative people. But he also met people who were positive, who wanted to learn, who wanted to achieve things. Mike decided to spend his time with the positive people. Almost immediately, Mike started to feel much better about himself. He developed a great attitude. He began to set goals.

Today is a Good Day!

I am happy to tell you that Mike now runs his own successful engineering company and has a wonderful family. One by one, he is accomplishing all of the goals he has set. When I asked Mike what happened to his high school friends, he told me, *"They still live in the same neighbourhood. They are still negative and they are still doing nothing with their lives"*. Mike added, *"I would never be where I am now if I had kept hanging out with those guys. I would be still be at the corner deli playing pinball"*.

Mike's story is a great reminder of the influence others have on our lives. And yet, sometimes we get in the habit of being with certain people – and we just don't think about the consequences.

Have you ever heard the axiom, *"Tell me who you hang out with and I'll tell you who you are?"* There is a lot of wisdom in that simple statement. Have you given much thought to how this principle has been molding and shaping your life?

Think back to when you were growing up. Do you remember how concerned your parents were about who you hung out with? Your mom or dad wanted to meet your friends and know all kinds of details about them. Why? Your parents knew you would be greatly influenced by your friends that you would tend to pick up some of their habits, and that you would probably do the things your friends were doing. Your parents were concerned for good reason. Let me give you an example. When I worked for a multinational company as a Project Manager, I started up a PMO. HP professional services organization was not very convinced over the urgency of creating a PMO, so I looked for some allies. Most of my Project Manager's colleagues criticized the lack of support from the executives, but they never proposed any action to be taken. One of them thought like me, he said, *"something can be done in terms of our behavior"*, in front of our executives. My colleague was my first ally in the PMO implementation project effort. *"Change is possible, and today is a good day"*, was our preferred sentence.

Positive Versus Negative People

Negative people always dwell on the negative. Their sentences are contagious, they continuously spew their verbal poison. In contrast, positive people "promote personal and professional growth", and they are very supportive. Positive people lift your spirits and are a gift to all of us.

Negative people always try to drag you down to their level. They hammer away at you with everything you can't do and all of the things that are impossible. They barrage you with gloomy statements about the lousy economy, the problems in their life, the problems soon to be in your life and the terrible prospects for the future. If you are lucky, they might even throw in a few words about their aches and pains.

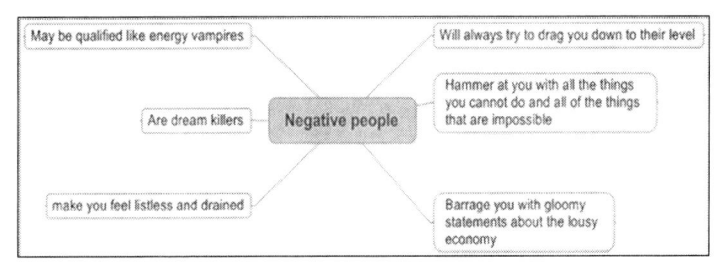

```
                                          Will always try to drag you down to their level
May be qualified like energy vampires
                                          Hammer at you with all the things
                                          you cannot do and all of the things
  Are dream killers    Negative people    that are impossible

                                          Barrage you with gloomy
  make you feel listless and drained      statements about the lousy
                                          economy
```

After listening to negative people, you feel listless and drained. Sometimes I identify people as *"dream killers"*. I call them *"energy vampires"* because they suck all the positive energy out of you. Have you ever been with a negative person, and felt as if that individual were physically taking energy from you? I think we have all had that experience at times. One thing is certain, spend time with negative people and their negative messages will wear you down.

Today is a Good Day!

On the other hand, how do you feel when you are around people who are positive, enthusiastic and supportive? You are energized and inspired. There is something truly amazing about positive people. They seem to have a positive energy that lights up a room. When you are around them, you start to pick up their attitude and you feel as if you have added strength to vigorously pursue your own goals.

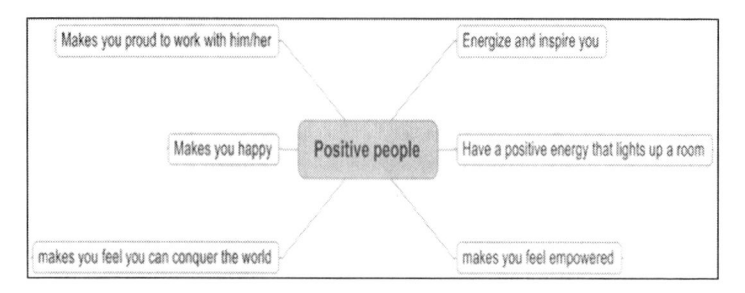

Positive people	
Makes you proud to work with him/her	Energize and inspire you
Makes you happy	Have a positive energy that lights up a room
makes you feel you can conquer the world	makes you feel empowered

When I think about positive people, immediately my friend Enrique Capella comes to mind. Whenever I speak with Enrique, I feel like I can conquer the world. Enrique is simply the most positive person you could ever meet.

I like to think of myself as a very positive person. On a scale of 1-10, with 10 being the most positive, I would probably give myself a 8.5. I would have to score Enrique a 14. He is just off the charts. He is always positive, always enthusiastic. And he gives a tremendous lift to everyone who crosses his path. His attitude inspires people to greatness. Can you see how your attitude might improve if you had a friend like John in your life?

Our mind tends to dwell upon whatever is repeated over and over. Unfortunately, the mind does not discriminate between messages that are good for us and those that are harmful. If we hear something often enough, we tend to believe it and act upon it. Just as a song repeated many times will cause

Associate With Positive Professionals

us to think about that song. So too, will repeated thoughts about success cause us to think about success.

So if we make sure we fill our mind with positive messages, we are going to be more positive and move forward boldly to achieve our goals. The more positive reinforcement the better off we are. And where can we get this positive reinforcement? Well, one way is to read motivational books. We can listen to motivational tapes and spend lots of time with positive people.

I believe that humans are like sponges. We *"soak up"* whatever people around us are saying. So, if we spend time with someone who is negative we soak up the negatives and it affects our attitude. Of course, the reverse is also true. When we hang around positive people we soak up the positive. We feel better and perform better. So, you must join positive people.

Analyze Your People

It is essential that you assess your friendships and professional colleagues from time to time, even those you have maintained for many years. Trust me; it's not a minor issue. Those who occupy your time have a significant impact on your most priceless possession, your mind.

Are you surrounding yourself with negative friends and colleagues and spending a lot of time with them in your leisure hours? If so, I am going to ask you to think about spending much less time with these people, or even no time at all with them.

Sounds harsh, doesn't it? After all, I'm suggesting that you limit or eliminate your involvement with some long-standing friends. You can think I'm cold or uncaring. You can also think we should try to help our negative friends and colleagues instead of dumping them. Nevertheless, I have

Today is a Good Day!

found in most cases, hanging around these negative friends or colleagues doesn't help them, and it doesn't help you, either. Everyone gets dragged down because most negative people don't want to change. They just want someone to listen to their tales of woe.

If you have a strong urge to spend time with negative people, ask yourself, *"Why am I choosing to be with these people?"* Consciously or unconsciously, you may be choosing to hold yourself back, to be less than you are capable of becoming.

By the way, I think it's wonderful to try to help someone overcome their negativity. But if you have been trying for several years and haven't been getting anywhere, may be it's time to move on.

Let me clarify one important thing. I'm not judging negative people as having any less worth than other people. I'm saying there are consequences if you spend time with people who are negative. What are the consequences? You will be less happy and less successful than you could be.

One of the things that has worked well for me in the projects I managed was when a discussion moved to a negative subject; I resisted the temptation to accuse the other person of being negative. Doing that will usually make things worse. Instead, gently shift the conversation to a more positive topic.

Remember, I am not asking you to disown your relatives or refuse to attend family functions. This is about limiting your contacts with negative relatives so you do not get dragged down to their level.

Do you Work With Positive People?

Every organization and every project includes some negative people. And some times you have to interact and work alongside these people. But don't go out of your way to spend time with these prophets of gloom and doom.

For example, if you frequently have lunch with negative people at work, stop having lunch with them. All they do is fill your mind with negativity. You can not perform at your best if you allow these people to dump their negative garbage into your mind. There is no need to be nasty or to tell them off. You should be able to find a diplomatic way of distancing yourself from this "poisonous" group.

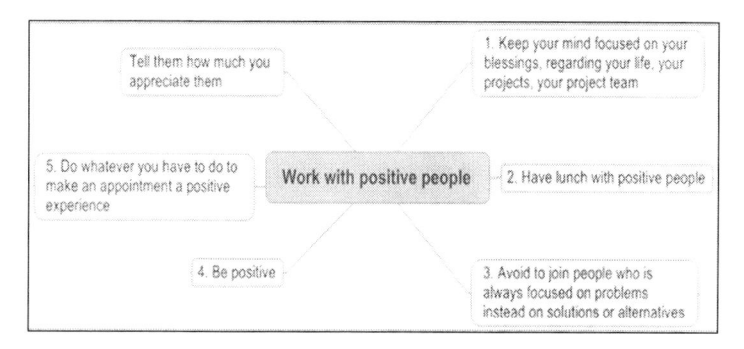

```
                 Tell them how much you                      1. Keep your mind focused on your
                 appreciate them                             blessings, regarding your life, your
                                                             projects, your project team

5. Do whatever you have to do to
   make an appointment a positive    Work with positive people    2. Have lunch with positive people
   experience

                 4. Be positive                              3. Avoid to join people who is
                                                             always focused on problems
                                                             instead on solutions or alternatives
```

Instead, take charge. Be proactive. Make a point to eat at your desk, take a client out to lunch or to sit at a different table in the cafeteria. Do whatever you have to do to make lunch a positive experience.

Make no mistake about it. Positive people are welcomed in any organization and negative people are hurting their chance for advancement. The business community is waking up to the fact that when it comes to productivity in the workplace, attitude is very important for project and business success. Today is a Good Day!

Choose Your Friends and Allies

As I said at the beginning of this chapter, *"Tell me who you hang out with and I'll tell you who you are."* If you are serious about obtaining a raise or a promotion at work, succeeding in your own business or improving yourself as a human being, then you have to start associating with people who can take you to the next level.

As you increase your associations with positive people, you will feel better about yourself and have renewed energy to achieve your goals. You will become a more positive, upbeat person, the kind of person others love to be around. I used to think it was important to associate with positive people and to limit involvement with negative people. Now, I believe it is essential if you want to be a high achiever and a happy individual.

So, surround yourself with positive people. They will lift you up the ladder of success.

Case study

After managing some important projects over the years, by 1993, I discovered PMI. I understood how important was to take care of my personal development as a Project Manager. Then, I discovered some different resources; one of them was the PMI Global Congress in the US. After thinking about the benefits for me as a Project Manager, for my peers, for my organization and for my customers, I was able to argue about the need to attend.

I hesitated at the beginning, but afterwards I proposed attending that Congress to my manager. As my performance had been very good the previous year, after listening to my arguments he accepted. However, he couldn't understand very well why a Spanish professional wanted to attend an international Congress. He said, "Americans are different from us. Spanish people are different". I answered, *"Thank God we have different cultures, but the point is to be ready to learn form anybody. And also it will be a great opportunity to meet new people and to share similar issues, and problems".*

It was a fantastic opportunity. At the conference, I discovered the huge power of networking. Over the years I still meet new positive people every year at PM International Congresses. It encourages me to think about how I am managing my projects, how to improve, and furthermore, it generates a lot of positivism and enthusiasm. When I return from a Congress I transmit that enthusiasm to my organizational team. That's great!

I strongly believe there is no border for project management. Positivism is the key for learning from projects and from people in organizations.

Summary

Remember the following ideas and best experiences about associating with positive professionals:

- Negative people are the ones who always dwell on the negative. Their sentences are contagious, they continuously spew their verbal poison. In contrast, positive people *"promote personal and professional growth"*. They are very supportive. Positive people lift your spirits and are a gift to all of us.

- It is essential that you assess your friendships and professional colleagues from time to time, even those you have maintained for many years

- Positive people are welcomed in any organization and negative people are hurting their chances for advancement. The business community is waking up to the fact that when it comes to productivity

Today is a Good Day!

in the workplace, attitude is very important for project and business success. Today is a Good Day!

CHAPTER 10

Grow Through Your Fears

> *Even the fear of death is nothing compared to the fear of not having lived authentically and fully.*
> —Frances Moore

I define myself as a human being always ready to learn something from somebody else, and I did this, managing projects, managing team members and other project stakeholders. I learned that if you want to be successful, you must be willing to be uncomfortable. To achieve your goals and realize your potential, you must be willing to do things that you are afraid to do. That is how you develop your potential. I encourage all Project Managers I work with, to follow this principle.

It sounds so simple, but it's not. What most people do when they face a frightening or new situation is to back away from the fear. They don't take action. That's what I did for many years of my life. I strongly believe that isn't the correct strategy. Tell me about a successful professional or Project Manager and I will show you someone who confronts his or her fears and takes action. I will tell you about somebody who never looses his/her courage.

Assess Your Fears

Many professionals I asked, *"have you ever been afraid or anxious before trying a new or challenging activity or project?"* They answered me, *"yes I was"*. Most of them told me that many times their fear stopped them from taking action. Sometimes people are paralyzed by fear.

I can remember the first project I managed. I was 27 years old, without any experience as a project leader. I had to share with a big group of Executives, the project mission, objectives, scope, initial risks and so on. I was very nervous .It was the first time I was in front of a big group of people and they were my project stakeholders. When I started my presentation, for a few seconds I was paralyzed, my mind stopped functioning. However, as soon as I started to talk, I was feeling more and more comfortable."

I learned those reactions are part of us as human beings. Of course, every individual has a different fear threshold. What frightens one person to death might have little impact on another. For example, some years ago, for me, to speak in public or start a new business was scary. Others might be fearful over asking someone for directions or for deadlines. Regardless of how trivial or silly you believe your fears may be, this lesson applies to you.

When I talk about fear, I am talking about those challenges that stand in the way of your personal and professional growth. These are the things that scare you, but which you know are necessary if you are going to get what you want in life.

The Comfortable Zone

Every time you step out of your comfort zone you will be gripped by fear and anxiety. Each of us has a comfort zone; a zone of behavior that is natural and familiar to us and where we feel comfortable and safe.

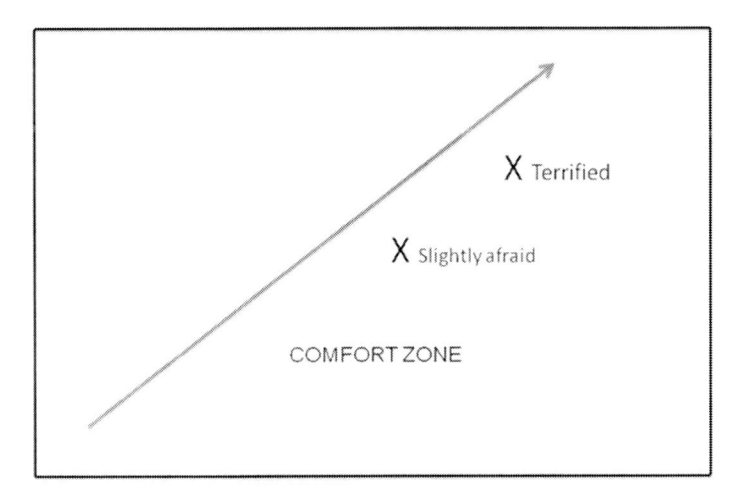

```
                                    ↗
                        X Terrified
                   X Slightly afraid

            COMFORT ZONE
```

The activities and situations that lie inside the comfort area are non-threatening and familiar. They are a routine part of your daily life, the things you can do with no sweat. In this category, there are tasks such as speaking to your friends or professional colleagues, customers, or filling out the daily paperwork at your job.

However, you as Project Manager face experiences or challenges that are outside your comfort zone. These are represented by the *"X's"* in the diagram above. The farther the *"X"* is from the comfort zone, the more afraid you are to participate in that activity.

When faced with something outside your comfort zone, you suddenly feel nervous. Your palms become sweaty, and your heart pounds. You begin to wonder, *"Will I able to handle it? Will others laugh at me? What will my colleagues and customers say?"*

As you look at the diagram above, what does the *"X"* represent for you? In other words, what fear is holding you back from reaching the next level of success or fulfilment in your life?

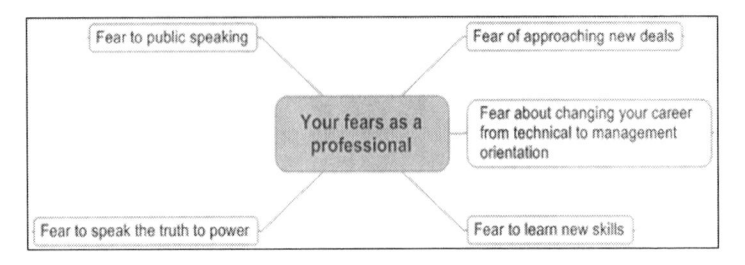

```
┌─────────────────────────────────────────────────────────────┐
│  Fear to public speaking          Fear of approaching new deals │
│                                                             │
│                    ┌──────────────┐  Fear about changing your career │
│                    │ Your fears as a │  from technical to management │
│                    │  professional  │  orientation                │
│                    └──────────────┘                         │
│                                                             │
│  Fear to speak the truth to power    Fear to learn new skills │
└─────────────────────────────────────────────────────────────┘
```

Whatever that *"X"* represents for you, just be honest and admit it. My guess is that thousands, if not millions, of people have the very same fears you have. In fact, let's take a closer look at what most Project Managers are afraid of.

The Most Common Fears

I have asked many Project Manager professionals about their most common fears as Project Managers, managing projects in organizations. I heard from professionals from different countries and cultures. The same answers popped up over and over. Here are some of the most common fears they identify:

1. Public speaking.
2. Saying No.
3. Changing jobs and starting a new business.
4. Telling managers and or executives *"negative news"*.
5. Talking to "upper managers".
6. Fear of failure.

Grow Through Your Fears

```
                         The most common
                         fears from Project
                         professionals

6. Fear of failure                            1. Public speaking
5. Talking to "Upper Managers"                2. Saying NO
4. Telling Managers or executives             3. Changing jobs or starting a new business
   "negative news"
```

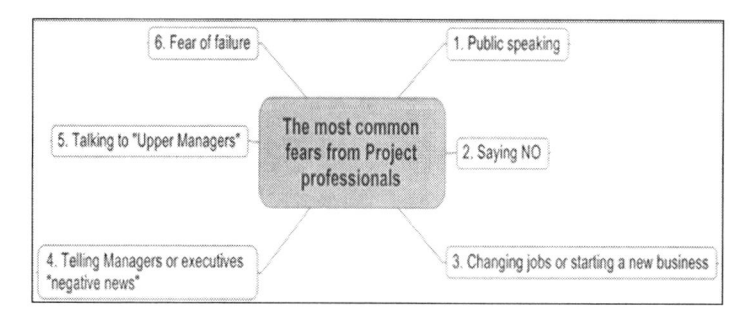

Are you surprised by any of the fears on this list? I believe the overwhelming majority of project professionals experience these fears at some point in their projects. And if you have some fears that are not on the list don't worry about it, TODAY IS A GOOD DAY! You are stronger than any of your fears and you can overcome them.

Backing Away From Your Fears

When confronted with an anxiety-producing event, most people will retreat to avoid the fear and anxiety. That's what I used to do. You see, backing away does relieve the fear and anxiety that would have resulted if you followed through with the activity. For example, if someone asks you to make a presentation within your organization, and you decline, you save yourself the sleepless nights you would have had worrying about it and the nervousness you would have experienced in the days leading up to the presentation.

In fact, I have found that the only benefit you receive by retreating is a momentary avoidance of anxiety. Think about it for a moment. Do you know of any other benefits that people receive when they refuse to confront their fears? Nobody has been able to tell me any additional benefits. And that's because I believe there are none.

The Price You Pay

Now, I want you to seriously consider the price you pay when you back away from those fears that are standing in the way of your growth. Here is what happens:

- Your self-esteem is lowered.
- You feel powerless and frustrated.
- You sabotage your success.
- You lead an uneventful, boring life.

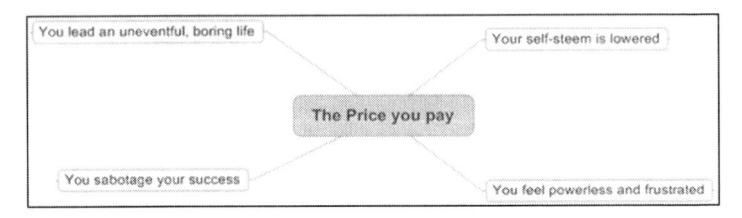

```
┌─────────────────────────────────────────────────────────────────┐
│ ┌──────────────────────────┐         ┌────────────────────────┐ │
│ │ You lead an uneventful,  │         │ Your self-steem is     │ │
│ │ boring life              │         │ lowered                │ │
│ └──────────────────────────┘         └────────────────────────┘ │
│                    ┌────────────────────────┐                   │
│                    │   The Price you pay    │                   │
│                    └────────────────────────┘                   │
│ ┌──────────────────────────┐         ┌────────────────────────┐ │
│ │ You sabotage your success│         │ You feel powerless     │ │
│ │                          │         │ and frustrated         │ │
│ └──────────────────────────┘         └────────────────────────┘ │
└─────────────────────────────────────────────────────────────────┘
```

Many of us are willing to pay this dear price, simply to avoid temporary discomfort and possible ridicule from others. I think it's not good. In the long run, retreating is not the best way to handle your problem. You will never be highly successful or develop your talents to the fullest unless you are willing to confront your fears.

Strategy

When I was in high school, I was pretty shy and didn't feel very good about myself. But I was never rejected when it came to asking someone for a date. If you looked at me now, you would probably think, "He is not bad looking, but he isn't Robert Redford either".

My strategy was quite simple. I never asked anyone out on a date. You see, I was not going to let anyone reject me. And what did I accomplish? I felt horrible about myself. I knew that

I had *"wimped out"*. I felt powerless, and as you can imagine, I didn't have a full social calendar. I was sabotaging my success.

Because I refused to face my fear, I remained in the background while most of my friends and classmates went out on dates. How do you think that made me feel? Pretty lousy, just as you would expect. In case you are wondering, I did have a few dates during that period of my life, but only when other people arranged them. I was not going to allow anyone to say *"NO"* to me. In reality, I was saying *"NO"* to myself.

Can you see how my strategy of backing away from my fears worked against me? Now it is true that if I had asked some people for a date in high school, a few of them might have said *"NO"*. But you know what? I wouldn't have died. I could have asked another person and another and eventually I would have gotten a *"YES"*.

It was not until college that I began to take some "baby steps" to confront my fear of rejection. Little by little, I gained more confidence. And while I was studying Computer Science I had the good fortune to meet Rose, who I have been married to for 23 years.

That same situation happened during my career as a Project Manager. I had some negative managers and sponsors when managing projects in organizations. However, I gained more and more confidence at the moment I was engaged in multinational projects and activities. I felt much more recognized and satisfied. That was when I discovered the great power of my positive behavior dealing with people. That insight changed my life completely. I started to talk to my executives and project sponsors as human beings first, and then as professionals. I lost some of my fears in front of them. Little by little I started to feel more comfortable. My fears were there but I felt more and more comfortable dealing with them. It was really great.

A New Life

I am no different from you. I have fears just as you do. And when I look back at the first 30 years of my life, you know what I see? I see someone who achieved some degree of success as an engineer. But I also see someone who was shy, insecure, scared and self-conscious. Does that sound to you like someone who is a motivational professional?

What turned my life around and improved it a million-fold, is when I learned to confront my fears and take action. I realized after years of frustration and disappointments that hiding from my fears was not getting me anywhere and it never would.

Of course, I wouldn't have confronted my fears if I had not first developed a positive attitude. A *"can-do"* attitude provided me with the extra push I needed to take action. When you believe you can do something you have the courage to move forward despite being fearful.

Armed with a great attitude, I decided to become a participant in life and to explore my potential, even though I was scared. From the very beginning, I felt so much better about myself. I had taken control of my life and all sorts of possibilities opened up for me.

Are you beginning to see the incredible rewards you can receive when you are willing to develop a positive attitude and confront your fears?

Re-frame the Situation

If I could give you a way to confront uncomfortable situations without fear or anxiety, you would be ecstatic and eternally grateful. Sorry, but there is no such magical solution. I can't wave a magic wand and take away your fears.

How then can you muster the courage to do those things that you fear, but which are necessary for your success and growth?

The next time you face a scary situation, I suggest you take a different outlook. Most people start thinking, *"I will not be able to do this well and other people may laugh at me or reject me"*. They get hung up over how well they are going to perform. Because of these worries they decide to retreat. While you should always go in with a positive attitude and prepare beforehand to the extent possible, don't be overly concerned with the result.

Consider yourself an immediate winner when you take the step and do the thing you fear. That is right. You are a winner just by entering the arena and participating, regardless of the result.

Moving Forward

For example, let's assume you are afraid to speak in public, but you confront your fear and do it anyway. The moment you get up and speak before the audience, you are a winner. Your knees may be shaking and your voice may be quivering. But that doesn't matter. You faced your fear and accepted the challenge. Congratulations are in order. The likely result is that your self-esteem will be enhanced and you will feel exhilarated.

Case Study

If we transported ourselves back in time, to, say, 1700 A.D., we would likely find that the medicine men and women in traditional cultures, were keen observers of nature, able to interpret weather patterns, the behavior of birds, animals and other things important to their survival. After all, they spent their entire lives in close interaction with the natural landscape

Today is a Good Day!

and were gifted with wisdom and a deep understanding of the cycles of nature.

Those men and women were really tribal psychiatrists and the caretakers of ancient medicine knowledge. As such, they were called upon to counsel, and their words were regarded carefully. In some instances, the shaman would supervise the vision quests of young people who were on the cusp of adulthood, and were about to embark upon adult lives as valuable members of society.

While on the vision quest, these young people were seeking the guidance and aid of spirit allies. The shaman would interpret the results of the quest, and give meaning to things that were otherwise difficult to understand. The vision quest was also a means of facing fear, real or imagined, because it took great courage to go off alone, to fast, and to endure various perils. In effect, the quest was a means of confronting fear; while the shaman was there to advise and direct the participant, and to help the person overcome fear of the unknown and of the spirit world.

To face and to confront one's fears, or other obstacles in life, is to gain an increasing measure of freedom. The result is that external influences or internal habits have less ability to shape your life. You become increasingly open to change, to a willingness to move in directions which change your circumstances for the better. In other words, fear loses its grip on your behavior patterns and on the way you live your life and the projects you manage.

Summary

- What most people do when they face a frightening or new situation is backing away from the fear. They do not take action. It is what I did for many years of my life. I strongly believe that is not the correct strategy.

Grow Through Your Fears

- When I talk about fear, I am talking about those challenges that stand in the way of your personal and professional growth. These are the things that scare you, but which you know are necessary if you are going to get what you want in life.

- When confronted with an anxiety-producing event, most people will retreat to avoid the fear and anxiety. That's what I used to do. You see, backing away does relieve the fear and anxiety that would have resulted if you followed through with the activity

- I am no different from you. I have fears just as you do. And when I look back at the first 30 years of my life, you know what I see? I see someone who achieved some degree of success as an engineer. But I also see someone who was shy, insecure, scared and self-conscious.

Today is a Good Day!

CHAPTER 11

Get Out There and Fail

You always pass failure on the way to success.
—Mickey Rooney (1920)

I can remember the words of my father a long time ago, "If you want to learn to do something well you must fail first." I believe failing is a good way to learn to do things better. Henry Ford said, *"Failure is only the opportunity to more intelligently begin again."* One of the lessons I learned as a Project Manager was that persistence is a key for project success.

I failed many times in my professional life but I was willing to keep failing, and keep failing until I succeeded. Alexander Graham Bell said, *"What is power is I can not say; all I know is that it exists and it becomes available only when a man is in that state of mind in which he knows exactly what he wants and is fully determined not to quit until he finds it".*

Projects are done by humans who make right and wrong decisions during the project life cycle. This is part of human behavior that many executives forget exists during project execution in organizations. How to learn from successes and failures characterizes a project learning organization. One of the obligations from executives is to plan with their Project Managers for doing retrospective analysis during the project life cycle for each project. If lessons learned sessions are not planned as part of the project plan, they never will happen.

Trying hard and making many attempts is known as commitment and persistence in a general sense. Taking into account that projects are uncertain endeavours, project teams and Project Managers achieve right or wrong results. Depending on the point of view of those teams and leaders, those results will be considered as failures or as opportunities to learn.

For instance, Thomas A. Edison *"failed"* more than five thousand times before inventing the incandescent light bulb. However, when some people talked to him about these *"failures"* he said, *"I did not fail, I discovered five thousand ways of how not to build an incandescent light bulb"*. A learning attitude is a good characteristic of the right project leader. Every day I can learn something in my project, and it doesn't matter if I learn from my people, from my customer, or from other project stakeholders. Good ideas or feedback can come from anywhere. The most important thing is how to move from failures to project success.

Some Years Ago

I can remember when I was a child and I learned how to ride a bicycle. You began with training wheels. Eventually, when those crutches were removed, keeping your balance became more difficult. I struggled to stay upright, maybe even falling a few times and scrapping myself. But I was learning an important early lesson about failure.

As you practiced, it is likely that one of your parents walked beside you shouting instructions, encouraging you and catching you as you lost balance. You were scared, but excited. You looked forward to the time when you would succeed, when you would at last ride free on your own. So you kept at it every day and eventually mastered the skill of riding a bike.

What contributed to your ultimate success in learning how to ride your bike?

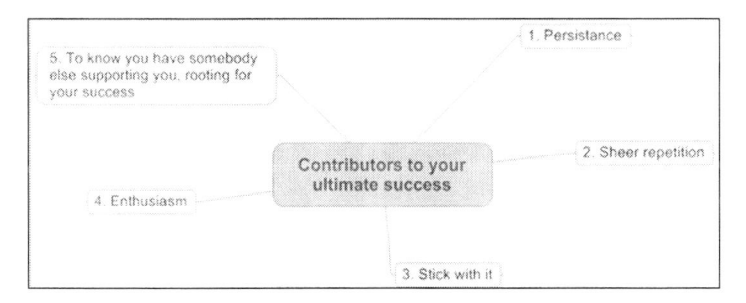

```
Contributors to your ultimate success:
1. Persistence
2. Sheer repetition
3. Stick with it
4. Enthusiasm
5. To know you have somebody else supporting you, rooting for your success
```

Well, persistence and sheer repetition, certainly. You were going to stick with it no matter how long it took. It also helped that you were enthusiastic about what you set out to achieve, that you could hardly wait to reach your goal. And finally, let's not underestimate the impact of positive encouragement. You always knew your parents were in your corner, supporting you, and rooting for your success.

As a six-year-old learning to ride your bike, you were optimistic, thrilled, and eager to meet the challenge. You couldn't wait to try again. You knew you would master it eventually. But that was a long time ago.

Yesterday and Today

Now let's examine how most adults approach the development of new skills. Let's assume we asked a group of adults to learn a new software program or to switch to another position in the company. How would most respond? They would try to avoid it, they would complain, they would make excuses why they shouldn't have to do it, they would doubt their abilities, and they would be afraid.

Today is a Good Day!

As adults, most of us become a lot more concerned about the opinions of others, often hesitating because people may laugh at us or criticize us. At the age of six, we knew we had to fall off the bike and get back on to learn a new skill. Falling off the bike was not a *"bad"* thing. But as we got older, we started to perceive falling off as a bad thing, rather than an essential part of the process of achieving our goal.

It can be uncomfortable to try something new, perhaps even scary. But if you take your eyes off the goal and instead focus your attention on how others may be viewing you, then you are doing yourself a grave disservice. To develop a new skill or reach a meaningful target, you must be committed to doing what it takes to get there, even if it means putting up with negative feedback or falling on your face now and then.

Successful people have learned to *"fail"* their way to success. While they may not particularly enjoy their *"failures"*, they recognize them as a necessary part of the road to victory. After all, becoming proficient at any skill requires time, effort and discipline, and the willingness to persevere through whatever difficulties may arise. Persistence is the key.

The Greatest Mistake

When I ask you to name the best basketball player of all time, who comes to mind? I am guessing that many of you immediately thought of Michael Jordan. He gets my vote. Let me share with you this statistic – Michael Jordan has a career shooting percentage of 50%. In other words, half of the slots he took in his career were *"failures"*.

Of course, this principle is not limited to sports. We also know that show business starts and media personalities are no strangers to failure. I spent many years learning the basics of project management, and I still make some mistakes in the projects I am engaged in. However, I try to learn from them. I recognize my failures in front of my people. I am not Superman. I make mistakes every day. But at the end of the day, I try to summarize my mistakes and promising myself to make adjustments in planning the next day will be great.

That is, if you keep trying, keep developing yourself and keep making adjustments along the way, you are going to succeed. You simply need to get enough at-bats, go on enough auditions, and visit enough potential clients.

Every time I make I mistake managing a project I recognize it and say, *"I made a mistake, I will do all my best to correct my mistake, my apologies about that"*. It is the type of behavior I get across to my people and has been very helpful in my life as a Project Manager. So, the greatest mistake a good Project Manager that wrongs something is not recognizing and saying, "I made a mistake".

The Power of Persistence

I believe commitment is the essence of a learning attitude. The key to getting what you want is the willingness to do "whatever it takes" to accomplish your objective. What do I mean by this willingness? It is a mental attitude that says, *"if it takes five steps to reach my goal, I'll take those 5 steps, but if it takes 30 steps to reach my goal, I will be persistent and take those 30 steps"*.

On most occasions in the project field, you don't know how many steps you must take to reach your goal or to accomplish your deliverables. This doesn't matter. To succeed, all that's necessary is that you make a commitment to do whatever it takes, regardless of the number of steps or activities involved.

Persistent action follows commitment. You must first be committed to something before you'll persist to achieve it. Once you make a commitment to achieve your goal, then you will follow through with relentless determination and action until you attain the desired result.

The most difficult thing I've found, is how to convince the team about the big impact on business that commitment within projects has on organizations. When you make a commitment and you are willing to do whatever it takes, including the effort to communicate a clear, convincing, and compelling message, you begin to attract the people and circumstances necessary to accomplish your goal.

For instance, some years ago I was responsible for implementing a PMO (Project Management Office) at an IT multinational company. The manager of that organization did not believe in project management. My team consisted of junior people, without experience in project management. However, I achieved tangible results in a few months. The key for project success was that my vision *"turned from potential failure to possible success"* in my mind and it was an incredible engine. I spent a lot of time meeting project stakeholders and searching for project allies in the organization, meeting team members, explaining the purpose and objectives of the Project Office, and training them on project management basics.

The key for project success in this story was spending time explaining that everyone must be accountable and responsible for their tasks and activities. That means getting team members to commitment. Speaking the truth to all project stakeholders about the project was also a key. I created a need to learn the project life cycle, identifying right and wrong insights. I believe that once you commit yourself to something, you create a mental picture of what it would be like to achieve it. Then, your mind immediately goes to work, attracting events, circumstances and people that help bring your picture into reality.

It is important to realize, however, that this is not a quick process, and you need to be persistent and, most importantly, educate and support your people about persistence, I believe that is a must for project success. Usually project failures precede project success, especially if all project stakeholders are ready to learn, take time to reflect, and take action towards achieving desired results.

My Rules of Persistence

Therefore, today I have decided on some *"Laws of persistence for my projects and for myself"*:

1. **I will have no regrets.** I will follow my dreams to the fullest. With all my energy I will give it my complete will and effort, so that even if the desired result does not come about, I will have no regrets. I know I tried.

2. **I will activate my dreams through little actions.** Yard by yard, push by little push. I need not take massive action each day. But a little measurable step forward will bring me that much more nearer to my goal.

3. **I will live in the moment, not in the past and not too much into the future.** The full realization of the present will make the future come about on its own. Let me effuse all my energies on the present so that I don't rue the time lost when the clock ticks over.

4. **I will always keep my goals in sight.** I have written down my goals and I carry a copy with me in my wallet. I am making a daily habit of at least going through the list once. The other places I have kept the list, is as a wallpaper on my monitor. It's always in my face and I hope in my subconscious too.

Today is a Good Day!

5. **I realize that obstacles will come about.** I need to work around them. Goals are what lie behind all the stumbling blocks. If I can't vault over them then I will walk around them. It might take longer, but I will get around the block.

6. **I will only focus on one or two goals.** Focus is concentration on one single point. It's much easier to be persistent when we have clarity of a single goal. Too many goals dissipate our energies and loss of energy is always followed by loss of persistence.

7. **I will trust myself.** When others can do it so can I. I try out this mantra every day. I know all the power to achieve my goals lies within me. I only have to harness it.

8. **I will take a break.** I have to fill myself up with energy. After every slight success it's important to taste a reward. Just to chill out for a while and then get back on the job rejuvenated.

9. **I will be flexible.** Constant action sometimes demands inconstant methods. If a way is not working too well, I will try to find some other way to do it. Their always is more than one way to bell the cat.

10. **I will be patient.** What defeats persistence is time. Time is our greatest friend as well as our greatest enemy. Persistent action by its very extension, means overcoming an obstacle over time. So I have to make time my ally and trust that with patience I will complete my goal. If I can progress a little each day, I will have utilized time.

Get Out There and Fail

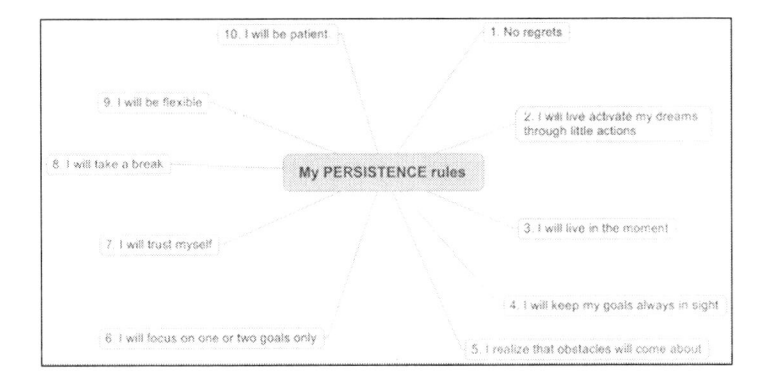

My PERSISTENCE rules:
1. No regrets
2. I will live activate my dreams through little actions
3. I will live in the moment
4. I will keep my goals always in sight
5. I realize that obstacles will come about
6. I will focus on one or two goals only
7. I will trust myself
8. I will take a break
9. I will be flexible
10. I will be patient

Never Give Up

By 2002, the IT manager of a Savings bank called our office to inquire about our project management services, as well as our products and publications. When we made the follow up call, he said that he was. "thinking about it and had not made a decision yet".

At the beginning, we called every week with no sale. When we called once a month there was no sale. Over a period of a few years, we continued to call this gentleman. We kept sending him quarterly newsletters and flyers. And all we had to show for it was one failure after another.

But in the autumn of 2004, a representative of his company called our office and we were hired to deliver a consulting and training program for more than 150 employees. When I met the IT manager in person he told me, *"I was impressed with your persistence. Someone from your office kept calling me for years and did not give up"*. Sure, we put up with years of failure. But it was all worth it when we made that sale.

Let me share with you another example. At the year 2007, I was elected as PMI Madrid Chapter President. PMI Madrid had about 250 members and had a lack of activities and no nice stories on its back. I am still dealing with a challenging situation. Some members of my board resigned. However, we now have a healthy financial situation and we have grown to 439 members. On the other hand, I lacked of volunteerism from this association (Volunteerism is very difficult in Spain). Moving forward is difficult but it is not impossible, TODAY IS A GOOD DAY. I will continue working hard and trying to motivate and encourage my team and our members.

Important Questions

If you are not getting the results you want or have been discouraged by failures, ask yourself these questions:

1. **Do I have an unrealistic timetable?** Maybe you expect to *"skip steps"* and succeed on a grand scale immediately. Success is usually achieved by climbing one step at time. And you do not always know how long it will take to advance to the next level. So, be patient with yourself – and resist the temptation to compare your progress with that of anyone else. You will advance faster than some, slower than others. Maintain a great attitude, take action, make adjustments, and the results will come.

2. **Am I truly committed?** It is essential for me that you be willing to do whatever it takes and that you banish any thought of giving up before you accomplish your objective. Of course, it is much easier to be committed when you love what you are doing. Therefore, go after those goals you are passionate about, and harbor no thought of quitting.

3. **Do I have too many discouraging influences?** Unsuccessful results can be frustrating. That's why

we need to surround ourselves with people who support and believe in us. If you hang around with negative people who are highly critical or who are doing very little in their own lives, your energy and enthusiasm will be drained. Therefore, develop a network of individuals to encourage and coach you toward success.

4. **Am I preparing to succeed?** Success in any endeavor requires thorough preparation. Are you taking steps to learn everything you can about accomplishing your goal? This means reading books, listening to tapes, taking courses and networking with highly successful people in your field. It might mean finding a mentor or getting a coach to work with you. Successful individuals are always sharpening their skills. Those getting unsuccessful outcomes often do the same things over and over without making necessary adjustments. So, be *"coachable"*. Accept the fact that you don't already know it all and find resources to keep you on track and moving forward.

5. **Am I truly willing to fail?** Face it, failure is inevitable. You will encounter defeat prior to succeeding. In our hearts, we know our most valuable lessons come from our failures. Failure is essential for growth. Look failure squarely in the face and see it as a natural part of the success process. Then, failure will lose its power over you. The truth is, when you are not afraid to fail, you are well on the way to success. Welcome failure as an unavoidable, yet vital component in the quest to achieve your goals.

Today is a Good Day!

Turning Failure into Success

Your failures are learning experiences that point out the adjustments you must make. Never try to hide from failure because that approach guarantees that you will take virtually no risks and achieve very little. S. Beverly Sills once remarked, *"You may be disappointed if you fail, but you are doomed if you do not try"*.

No, you will not close every sale. And you will not make money on every investment. Life is a series of wins and losses, even for the most successful. The winners in life know that you crawl before you walk and you walk before you run. And with each new goal comes a new set of failures. It is up to you whether you treat these disappointments as temporary setbacks and challenges to overcome, or as insurmountable obstacles.

If you make it your business to learn from every defeat and stay focused on the end result you wish to attain, failure will eventually lead you to success.

Case Study

The Maine Medicaid Claims System project is a case study of a project gone awry. The project was undertaken to switch from their legacy systems to a new web-based system to process Medicaid claims and facilitate HIPAA compliance (Health Insurance Portability and Accountability Act of 1996). As a result of the failed project, Maine is now the only state in the union not in compliance with HIPAA.

System problems led to many claims ending up in limbo, leading to hundreds of calls from health care practitioners, with nearly 300,000 patients being turned away, several dentists and therapists going out of business, and destroying Maine's finances and credit rating. So what went wrong?

Mistakes included the following:

- Deciding to develop an entire system from scratch using unproven technology, while other states built a front-end onto their legacy systems.
- Caving to pressure from management to meet tight deadlines with inadequate resources instead of pushing for a realistic plan to begin with.
- Failing to notice why other bidders, either didn't bid or came in way higher (a sign that the schedule was unrealistic).
- Hiring a vendor with no experience in developing Medicaid claims systems because they were the lowest bidder.
- Not having a Medicaid expert on the team, leading to errors in judgment.
- Underestimating the time needed to meet with subject matter experts.
- Competing with another major initiative (a department merger) for executives' attention and resources.
- Skipping project management basics (including piloting, adequate end-to-end testing, staff and user training, etc.) due to looming deadline pressures.
- Failing to stop, regroup, and analyze the risks
- Taking a "big bang" approach to cut over with no contingency or backup should something go wrong.

Management's response, of course, was to switch program managers, and issue stronger demands to have a smooth system. But none of the changes or demands made much of a difference. Consultants were brought in to prioritize the many problems, but still, the complexities proved too much. It wasn't until a Medicaid expert was brought in that things began to gel.

Like many project failures, it's easy to point to the project management (and certainly there are many shortcomings there in this case), but the organization must share the blame as well if it insists on unrealistic deadlines and leads by fear, (fear of shareholders, fear of competition, fear of management, etc.). None of these variables can make an unrealistic schedule more realistic.

It's really very simple. Either adequate resources must be committed, the expectations lowered, or a more piecemeal approach taken (or all three, if applicable). In any case, the schedule must be realistic and risks need to be managed.

Summary

- Projects are done by humans who make right and wrong decisions during the project life cycle. This is part of human behavior that many executives forget exists during project execution in organizations.

- Good ideas or feedback can come from anywhere. The most important thing is how to move from failures to project success

- It can be uncomfortable to try something new, perhaps even scary. But if you take your eyes off the goal and instead focus your attention on how others may be viewing you, then you are doing yourself a grave disservice

- Every time I make I mistake managing a project I recognize it say, *"I made a mistake, I will do all my best to correct my mistake, my apologies about that"*. It is the type of behavior I get across my people.

- I believe commitment is the essence of a learning attitude. The key to getting what you want is the willingness to do *"whatever it takes"* to accomplish your objective.

- It is important to realize, however, that this is not a quick process, and you need to be persistent and most important, educate and support your people about persistence, I believe that is a must for project success.

Today is a Good Day!

CHAPTER 12

Networking

*Everyone is born with genius, but most
people only keep it a few minutes.*
—Edgard Varese

The sooner you start creating a network, the faster you will progress in your career. When I joined PMI (Project Management Institute) in March 1993, it set off an incredible chain reaction, that would forever impact my professional life, and then my current business. Let me share with you what happened.

At the end of 1992, I attended project management training in France, organized by HP, the company I had worked for, almost fourteen years. In that training the teacher distributed to the attendees some project management articles. I knew about the existence of PMI (Project Management Institute) as a professional association. I asked my HP manager for permission to attend the PMI Global Congress in 1993. And after some discussions, he accepted my request. A huge window opened to me when I attended.

The first day of the Congress, I was a little bit frustrated because I was the only Spanish professional in attendance, and I was conscious that a lot of project management practitioners in Spain were not there. I attended a session called Global Forum being organized by Mr. David Pells and there I met most of professionals I would be in relationship some years later. I had the opportunity to distribute a lot of business cards. I collected many cards from colleagues from different countries and areas of expertise, and I really had a good time talking to and connecting with people.

That first event was very powerful for me. It motivated me and I understood the huge power of networking with people. Over the years I continued attending those annual Project Management Congresses. I had a developed a large network that is being increased every year. The power of networking is nothing short of awesome.

If given the choice, wouldn't you like to succeed sooner rather than later? Well, networking is a way to leverage your own efforts and accelerate the pace at which you get results. I strongly believe that the more solid relationships you build, the greater your opportunities are for success.

The Great Benefits of Networking

I believe that your success starts with you. However, it grows to higher levels as result of your associations and relationships with people. Simply put, you can not succeed on a grand scale all by yourself.

That is why networking is so important. Networking may be defined as the development of relationships with people for mutual benefit. I always took care of keeping my network alive. Some of the business benefits you can get as a project professional from networking activities are:

1. Generates new clients or business leads.
2. Increase business and professional opportunities.
3. Helps in finding the right people to fill critical positions or jobs.

4. Provides valuable information and resources.

5. Helps your professional relationships for personal and professional growth.

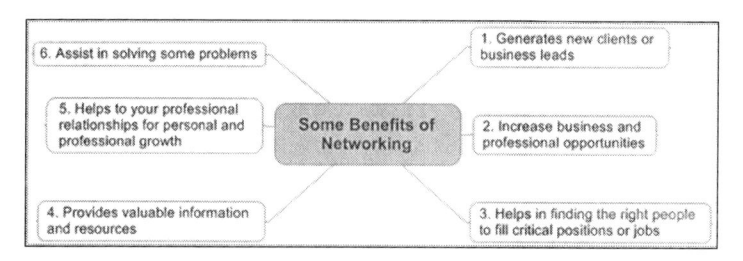

[Diagram: Some Benefits of Networking
1. Generates new clients or business leads
2. Increase business and professional opportunities
3. Helps in finding the right people to fill critical positions or jobs
4. Provides valuable information and resources
5. Helps to your professional relationships for personal and professional growth
6. Assist in solving some problems]

But, what can we do to enhance the effectiveness of our network? I found some productive techniques that have been very helpful for me. I have classified in different categories: Taking action, References, Communication and follow up.

Taking Action

1. **You must project a winning attitude.** When we talk about networking, attitude is a key for success. If you are positive and enthusiastic, people will want to spend time with you. They will want to help you. If you are gloomy and negative, people will avoid you, and they will hesitate to refer you to their colleagues.

2. **Be active in organizations and associations.** Effective networking and relationship building takes more than paying dues, putting your name in a directory and showing up for meetings. You must demonstrate that you will take the time and make the effort to contribute to the group. What kinds of things can you do? For starters, you can volunteer for committees or serve as an officer or member of the board of directors. The other

members will respect you when they see you roll up your sleeves and do some work. They will also learn about your people skills, your character, your values and last, but not least, your attitude will gladly be accepted.

Let me give you an example of Francisco. He wanted to build contacts within the electronics market, so he joined the Electronics Firms Association in Madrid in 1998. Francisco immediately began to attend the group's meetings. When they asked for volunteers for various projects, Francisco raised his hand. He got actively involved.

Within six months after joining, somebody approached him and said, *"We hear good things about you. You are a hard worker and very energetic. Would you like to join our board of directors?"* As you might guess, Francisco gladly accepted. And within a few months, he began to see a significant increase in his business. In early 1999, Francisco told me that well over 50% of his current business could be traced to people he met through the Electronics Association, proving that people can get big results in a short time by networking effectively.

3. **Serve others in your network.** Serving others is crucial to building and benefiting from your network. You should always be thinking, *"How can I serve others?"*, instead of, *"What's in it for me?"* If you come across as desperate or as a "taker" rather than a *"giver"*, you won't find people willing to help you. Going the extra mile for others is the best way to get the flow of good things coming back to you. How can you serve others in your

network? Start by referring business leads or potential customers. In addition, whenever you see an article or other information that might be of interest to someone in your network, forward the material to that person.

When I think of effective networkers, the first name that comes to mind is Jim De Piante, (US project professional). Jim works as a PM practitioner for a multinational company, He delivers presentations to project professionals at PM International Congresses and Events on soft skills, always transmitting his power, positivism and energy. I have referred many people to Jim. Why? He is a talented, service-oriented person who has gone out of his way to encourage me, and to help me to increase the power of networking.

Jim has put me in touch with people in his own network who are in a position to help me. He distributes his materials at his presentations. Jim is one of those people who just keeps giving, and giving, and giving. That's why people want to help Jim and that is one reason why his image, visibility and professionalism are growing more and more internationally.

Other powerful example is my colleague Michel Thiry, who is very active professionally in project management. He has a special charisma, that transmits interest in other people for networking purposes. He has increased his professional network very fast in the last years. How? By observing people at PM Congresses and inviting them to talk and to join his network, by having dinner together, by exchanging experiences, by finding ways to do business together, and by being proactive. His Valence Network is a real good example.

References

If you refer someone, make sure that the person mentions your name as the source of the referral. Be explicit. Let's assume you are about to refer John Smith to your graphic designer, Jane Jones. You might say to John, *"Give Jane a call, and please tell her that I referred you"*. In some instances, you may even call Jane and let her know that John Smith will be contacting her.

Then, the next time you see or speak to Jane, remember to ask if John called and how it turned out. You want to reinforce in Jane's mind that you are looking out for her and helping her to grow her business.

Be selective. Don't refer every person you meet. Respect the time of those in your network. Referring *"unqualified"* leads will reflect poorly on you. Ask yourself whether or not a particular referral is really going to be of value to your network partner. Keep in mind that the key is the quality, not quantity, of the leads you supply.

Communication

Be a good listener. Have you ever been speaking to someone who goes on and on about himself and his business and never takes a moment to ask about you? We have all run into the *"Me, me, me"* types and they are the last people you want to help. So, in your conversations, focus on drawing other people out. Let them talk about their careers and interests. In return, you will be perceived as caring, concerned and intelligent. You will eventually have your turn to talk about yourself.

Call people from time to time just because you care. How do you feel when someone calls you on the phone and says, *"Hey, I was just thinking about you and was wondering how you are doing?"* I'll bet you feel like a million bucks. If that's the case, why don't we make these calls more often? Every now and then, make it a point to call people in your network to

Networking

simply ask how they are doing and to offer your support and encouragement.

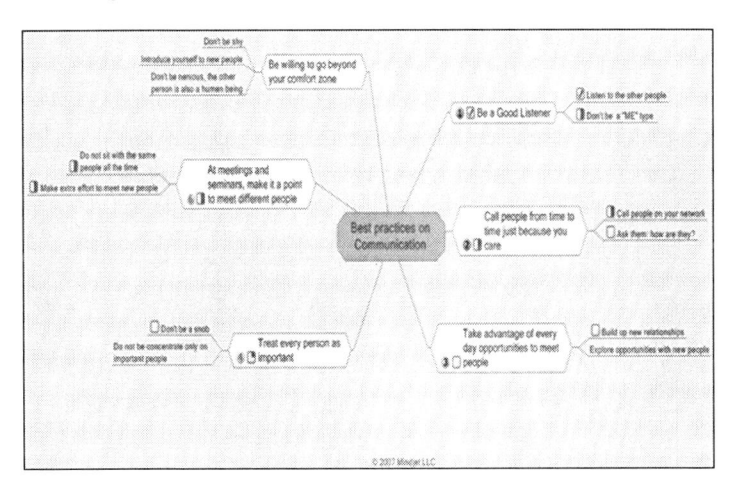

[Mind map figure: Best practices on Communication — Don't be shy (Introduce yourself to new people; Don't be nervous, the other person is also a human being); Be willing to go beyond your comfort zone; Be a Good Listener (Listen to the other people; Don't be a "ME" type); At meetings and seminars, make it a point to meet different people (Do not sit with the same people all the time; Make extra effort to meet new people); Call people from time to time just because you care (Call people on your network; Ask them: how are they?); Take advantage of every day opportunities to meet people (Build up new relationships; Explore opportunities with new people); Treat every person as important (Don't be a snob; Do not be concentrate only on important people). © 2007 Mindjet LLC]

That's right. Call just because you care and because that is the way you would like to be treated. Every December, I pick up the phone and call certain clients I haven't spoken with for a long time. Many of these people have not ordered anything from my company in years. My call is upbeat and my only agenda is to be friendly. I don't try to sell them anything. I appreciate the business they have given me in the past, and I just want to hear how they are doing, personally and professionally.

If business comes from these calls that's great. Year after year, I do get business as a result of making these calls. Someone will say, *"I need to order more of these services of Project management"*, or "Our Company is having a sales meeting in six months, and they may want you to do a presentation".

Please understand that this is not manipulation or a sales tactic on my part. I am not expecting these people to give me business. I really care about how they are doing. Business is simply a by-product of reconnecting with them.

Today is a Good Day!

Take advantage of everyday opportunities to meet people. You can make excellent contacts just about anywhere. You never know from what seed your next valuable relationship will sprout.

Treat every person as important, not just the *"influential"* ones. Don't be a snob. The person you meet (whether or not they are the boss) may have a friend or relative who can benefit from your product or service. So, when speaking to someone at a meeting or party, give that person your undivided attention.

And please promise me that you won't be one of those who gaze around looking for *"more important people"* to talk to. That really bugs me. You are talking with someone and then he notices someone out of the corner of his eye, someone he deems more important than you. So he stops listening to you, and abruptly breaks away to start a conversation with that other person. Don't do that! Treat every person you encounter with dignity and respect.

At meetings and seminars, make it a point to meet different people. Don't sit with the same group at every gathering. While it is great to talk with friends for part of the meeting, you will reap greater benefits if you make the extra effort to meet new faces. In 1996, I was in Washington, D.C., to attend a project management training. At lunch, instead of sitting with some friends from HP, I sat down at a table where I did not know anyone. Sitting at that table was a man named F.A. We struck up a conversation. His organization, conducts excellent training programs on soft skills for professionals.

It turned out that Frank is also a big believer that attitude is very important. Frank has become a good friend. I am sure glad I did not sit with my friends that day, as I would have missed out on a tremendous opportunity. Be willing to go beyond your comfort zone. For instance, if you have the urge to introduce yourself to someone, Do it! You might hesitate, thinking that the person is too important or too busy to speak with you. Even if you are nervous, force yourself to move forward and make contact. You will get more comfortable as time goes on.

Ask for what you want. By helping others, you have now earned the right to request assistance yourself. Don't be shy. As long as you have done your best to serve those in your network, they will be more than willing to return the favor.

Follow-up

Send a prompt note after meeting someone for the first time. Let's say you attend a dinner and make a new contact. Send a short note as soon as possible explaining how much you enjoyed meeting and talking with him or her. Enclose some of your own materials and perhaps include information that might be of interest to this person. Ask if there is anything you can do to assist this individual. Be sure to send the note within 48 hours after your initial meeting so that it is received while you are still fresh in your contact's mind.

Acknowledge powerful presentations or articles. If you hear an interesting presentation or read a great article, send a note to the speaker or writer and tell him or her how much you enjoyed and learned from their message. One person in a hundred will take the time to do this, be the one who does. I am not saying that speakers and writers often have developed a huge network of people covering a variety of industries, a network you can tap into.

When you receive a reference or helpful written materials, ALWAYS send a thank you note or call to express your appreciation. Follow this suggestion only if you want to receive more references and more useful information. If you don't acknowledge that person sufficiently, he or she will be much less likely to assist you in the future.

Send congratulatory cards and letters. If someone in your network gets a promotion, award or celebrates some other occasion write a short note of congratulations. Everyone loves to be recognized, yet very few people take the time to do this. Being thoughtful in this manner can only make you stand out. It is also appropriate to send a card or memorial gift when a family member dies.

Building Your Network

The networking suggestions offered above are merely the tip of the iceberg. You should be able to come up with several ideas of your own. How? By going to your library or bookstore and seeking out the many excellent books on networking, and by noticing what other people are doing and adapting their ideas in a way that suits you.

Remember that networks are built over time and that significant results usually don't show up immediately. Passion, persistence and patience must be cultivated if you want to increase your network. Build a solid foundation of relationships and continue to expand and strengthen them. You will have to put in a lot before you begin reaping the big rewards.

Finally, great networking skills are not a substitute for being excellent in your field. You might be a terrific person, but if you are not talented at what you do, and constantly learning and improving, your efforts will yield disappointing results.

Now move forward. Select a few of these networking techniques and implement them right away. Get to work serving and improving your network. Then you will truly have an army of troops working to help you succeed. Today is a Good Day!

Avoid Networking Gaffes

I have some best practices to share with you:

1. **Choose your flavorite.** Don't jump at every offer to join a professional network.

2. **Understand site culture and rules.** Before contacting the colleagues of the friend who invited you to network, get to know the group's culture.

3. **Hone your profile.** Review your profile and underline those characteristics that are important for you to show to other people.

4. **Don't be pushy.** Most of professional networks don't like arrogant people.

5. **Do what you say you will.** Practice authenticity and integrity.

6. **Prepare for face to face introductions.** A face to face meeting requires you to respond without the time afforded by email to craft your message. Know what you want for a meeting.

7. **Help yourself by helping others.** Networking is reciprocal, so do unto others as you'd want done to you. If you are able to help people, they will be more likely to remember you and return the favor.

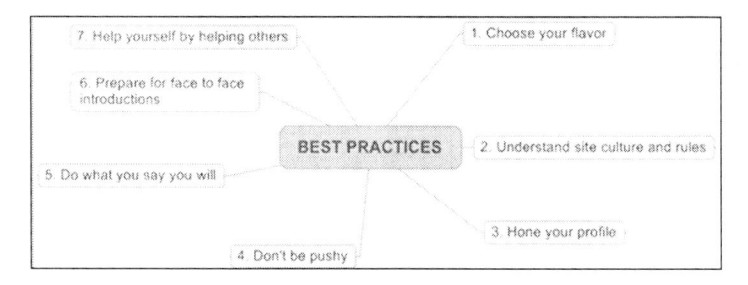

```
7. Help yourself by helping others        1. Choose your flavor
   6. Prepare for face to face
      introductions
                        BEST PRACTICES   2. Understand site culture and rules
5. Do what you say you will
                                          3. Hone your profile
                  4. Don't be pushy
```

Create a Personal Networking Plan

Professional networking is also a project, so you must prepare a plan for that project. It is critical that you clearly identify your network contacts, develop a personalized networking plan and build an administrative process to manage it all. It is very important to ask your network contacts for their help, not for a job. People are delighted to help, but few will have a job to offer you.

1. **First level contacts:** These are the hottest prospects and people you know best, current and past colleagues and managers, vendors, consultants, and recruiters, with whom you have an established relationship, Your initial contact will likely be via phone, for instance a quick call announcing you are in the job market and would appreciate advice, assistance, recommendations or referrals.

   At end of each conversation, tell your contacts you'd like to send them a resume to have on file and ask if they prefer mail, fax or email. Immediately forward your resume with a brief, friendly cover letter, thanking them for any help they can offer and mentioning the positions and industries in which you are interested.

   If you have not heard back from contacts within three weeks, call and inquire if they are reviewed your resume and if they have any recommendations.

2. **Second level contacts:** These are people you know casually. Your initial contact will most likely be 50% by phone and 50% by mail or email, depending on how comfortable you are in these relationships and how easy it is to connect with

each individual. Whenever possible, it is best if the initial contact is a phone call, allowing you to establish a more personal relationship. If you have called a contact, follow up immediately by sending a resume. If you have not heard back from contacts within three weeks, call or email them and inquire if they have reviewed your resume and if they have any recommendations.

Once you have developed your list of contacts and determined how to connect with each individual, set up a system to track all your calls, contacts and follow-up commitments.

Case Study

Since 1993 I have never lost the opportunity to attend the international project management Congresses every year. Thanks to knowing people from Malasya, Japan, India, USA, Costa Rica, Panamá, Argentina, Peru, Chile, Mexico, Colombia, Venezuela, Uruguay, Cuba, Brasil, Morocco, Malta, France, Holland, Belgium, Switzerland, Sweden, Norway, Denmark, Russia, Luxemburg, Italia, Greece, Portugal, UK, Ireland, Arabia, Australia, Rumania, Hungary, Croatia, and Slovenia.

When I started my adventure in 1993, delivering a presentation in English, first of all was a challenge for me, because my English level was very poor. Secondly, it was a big responsibility because I represented my organization internationally and I needed to do my best at all times. And lastly, it was a special effort being done on top of. However, I discovered the internal human power of enthusiasm, which encouraged me to move forward and improve my professional skills year by year. I met great people who advised me very positively. I knew people who understood the big power of networking, the huge ability to connect with people sharing experiences, failures, successes, great adventures, and great projects.

I learned that good networking also requires discipline from the professional, because you can add professionals to your network but you must sustain them. It hasn't been easy for me but it wasn't impossible. I maintain my contacts database as active as possible, I have lunch with different colleagues every month and I keep in touch periodically with most of my network colleagues.

It's been very helpful for me when I have managed international projects. To know people and have friends worldwide has been very beneficial. It is because I take care of my network month by month. I am a member of some project management networks. These communities have the objective of facilitating the exchange of experiences related to areas of knowledge of *"Project Management"* with the aim of promoting personal and professional growth of the individual.

Summary

Networking is very powerful for you as a Project Manager. Remember some of the best practices:

- The sooner you start creating a network, the faster you will progress in your career.

- Your success starts with you. However, it grows to higher levels as a result of your associations and relationships with people.

- Be a good listener.

- Call people from time to time just because you care.

- Treat every person as important, not just the "influential roles".

Networking

- Send a prompt note after meeting someone for the first time.

Today is a Good Day!

CHAPTER *13*

Conclusions

If you change your attitude, I really believe you will be able to change your life as a Project Manager, managing projects. If I would say that my last five years were a string of successes I would be lying. I had some failures and some successes too along the way. However, the real truth is that I felt happier when I opened my project window and I discovered the huge opportunity I had to learn more from all my colleagues and different project stakeholders.

The project of creating a new company was hard, stressful and challenging, but it has been worth it. All the ideas and suggestions I have discussed in this book are the fruit of my learning as a Project Manager in organizations, serving my customers and sharing and learning from other colleagues. If I had to explain my great discovery I would say, *"project management is my passion. Apply courage to your projects, control your fears and try to feel uncomfortable for a period of time. I am sure you will get something in return."* I try to apply the three P's every day (Passion, Persistence and Patience). Some days are more successful than others, but every day is a Good Day. Many times we are not aware of the blessings we have and we must say thank you for that.

Take Control of Your Projects

Thank you very much for reading this book. If you are reading my conclusions it's because you started to apply the three P's. I assume you want to increase and develop your professional potential as a Project Manager or as a Project Sponsor. This book wants to be your first step for helping you to live your projects in the way you want to live them. When you are focused on these ideas and thoughts and take action to implement them, you will be on the way to creating some exciting breakthroughs in your projects.

In any case, I sometimes feel like a UFO (Unknown Flying Object) in the project world in my country (Spain). It is very rare to find an individual who applies the ideas suggested in this book on a daily basis. And I will say it again, *"It is very difficult, but it is not impossible"*. However, it is very rare the professional who consistently maintains a positive attitude, knowing that his thoughts will become his reality. You only need to observe your colleague's faces early in the morning and see what happens.

It is also rare for the professional who watches the words he or she uses, knowing that he/she is programming his/her mind for success, mediocrity or failure. It is rare the professional who has the guts to confront his/her fears, because that is where his/her potential will be developed, by doing things he or she is afraid to do.

It is rare for the professional who looks for the silver lining in every dark cloud. Finally, it is very rare for the professional who makes a commitment, follows through with a positive attitude and has the passion, persistence and patience to get the job done.

I encourage you to be one of those rare professionals. Please join my project professionals *Rare's Club*. You have the potential to become more than you ever dreamed. You have

Conclusions

greatness within you, and your attitude is one of the keys to unlock that potential. In changing my attitude, I changed many things in my profession, yet I still have many things to change regarding my attitude. And if a better attitude changed many things in my professional and personal life, it can do the same for you.

I strongly believe we have a choice every day regarding the attitude we will embrace for that day. We can't change our past, our project issues, and we can't change the fact that people will act in a certain way. We can't change the inevitable. The only thing we can do is play on the one string we have, and that is our attitude. I am convinced that life is 10% of what happens to me and 90% of how I react to it. We are responsible for our attitude. So then, please take control of your attitude in front of your projects, with your team members, sponsors and other project stakeholders.

Please stop doing for a moment and think about who you are, where you are as a project professional, and where you want to go; and then prepare your plan to improve your attitude. You can do it, so do it now.

Move forward and believe in yourself.

Today is a good day…!

Today is a Good Day!

About the Author

ALFONSO BUCERO, DEA, PMP, is now an independent project management consultant and speaker. He is founder, partner, and director of BUCERO PM Consulting in Spain (http://www.abucero.com). He was managing director at IIL, Spain for two and a half years in Madrid while also serving as a project management consultant and trainer. His background was that of a Project Manager at Hewlett-Packard Consulting, where he developed and managed the PMO implementation whose purpose was the continuous improvement of project management discipline across the organization. During his thirteen years at HP, he managed various customer, infrastructure, development, and change management projects. He spent his last two years at HP selling and implementing the project office and convincing upper managers about the advantages of project sponsorship.

Today is a Good Day!

At IIL, Bucero consulted and trained executives, explaining the need for and advantages of good project sponsorship. He has a B.S. in computer science engineering. He is an international project management assessor of the International Project Management Association. He has a D.E.P (Diploma de Estudios Avanzados). He is preparing his Thesis about the Project Management Behavior implementing a Project Office. He was former president of the Project Management Institute's Barcelona chapter. Currently he is the President of PMI Madrid Chapter, and is a frequent contributor to international project management conferences and project office workshops. He is the contributing editor of the "*Crossing Borders*" column in the Project Management Institute's *PM Network* magazine. He is the author of the Spanish book, *Project Management: A New Vision*, contributed a chapter to Englund, Graham, and Dinsmore's *Creating the Project Office,* and he coauthored with Randall L. Englund, the book *Project Sponsorship – Achieving Management Commitment for Project Success.*

The author brings a practical slant to this book—the point of view of a professional who has been through it all. He is a practitioner of project management who practices his passion, persistence, and patience for implementing project management in organizations. He brings forth lessons that come from moving between observation, reflection and action. He put effort into thinking about attitude in project management, leveraging best practices or inventing new ones, applying or experimenting in their field of practice, observing and documenting the results, making modifications, and then doing it all again. This book is a product of those lessons.

Did you like this book?

If you enjoyed this book, you will find more interesting books at

www.MMPubs.com

Please take the time to let us know how you liked this book. Even short reviews of 2-3 sentences can be helpful and may be used in our marketing materials. If you take the time to post a review for this book on Amazon.com, let us know when the review is posted and you will receive a free audiobook or ebook from our catalog. Simply email the link to the review once it is live on Amazon.com, with your name, and your mailing address—send the email to orders@mmpubs.com with the subject line "Book Review Posted on Amazon."

If you have questions about this book, our customer loyalty program, or our review rewards program, please contact us at info@mmpubs.com.

Multi-Media Publications Inc.

Managing Smaller Projects: A Practical Approach

So called "small projects" can have potentially alarming consequences if they go wrong, but their control is often left to chance. The solution is to adapt tried and tested project management techniques.

This book provides a low overhead, highly practical way of looking after small projects. It covers all the essential skills: from project start-up, to managing risk, quality and change, through to controlling the project with a simple control system. It cuts through the jargon of project management and provides a framework that is as useful to those lacking formal training, as it is to those who are skilled project managers and want to control smaller projects without the burden of bureaucracy.

Read this best-selling book from the U.K., now making its North American debut. *IEE Engineering Management* praises the book, noting that "Simply put, this book is about helping people control smaller projects in a logical and effective way, and making the process run smoothly, and is indeed a success in achieving that goal."

Available in paperback format. Order from your local bookseller or directly from the publisher at

http://www.mmpubs.com/msp

Winston Churchill: The Agile Project Manager

Today's pace of change has reached unprecedented levels only seen in times of war. As a result, project management has changed accordingly with the pressure to deliver and make things count quickly. This recording looks back at a period of incredible change and mines lessons for Project Managers today.

In May 1940, the United Kingdom (UK) was facing a dire situation, an imminent invasion. As the evacuation of Dunkirk unfolded, the scale of the disaster became apparent. The army abandoned 90% of its equipment, the RAF fighter losses were deplorable, and over 200 ships were lost.

Winston Churchill, one of the greatest leaders of the 20th century, was swept into power. With depleted forces and no organized defense, the situation required a near miracle. Churchill had to mobilize quickly and act with agility to assemble a defense. He had to make the right investment choices, deploy resources, and deliver a complete project in a fraction of the time. This recording looks at Churchill as an agile Project Manger, turning a disastrous situation into an unexpected victory.

ISBN: 1-895186-50-1   (Audio CD)
ISBN: 1-897326-38-6   (DVD)

http://www.mmpubs.com

Managing Agile Projects

Are you being asked to manage a project with unclear requirements, high levels of change, or a team using Extreme Programming or other Agile Methods?

If you are a project manager or team leader who is interested in learning the secrets of successfully controlling and delivering agile projects, then this is the book for you.

From learning how agile projects are different from traditional projects, to detailed guidance on a number of agile management techniques and how to introduce them onto your own projects, this book has the insider secrets from some of the industry experts – the visionaries who developed the agile methodologies in the first place.

ISBN: 1-895186-11-0 (paperback)
ISBN: 1-895186-12-9 (PDF ebook)

Also available in ebook formats. Order from your local bookseller, Amazon.com, or directly from the publisher at

http://www.mmpubs.com

Lessons from the Ranch for Today's Business Manager

The lure of the open plain, boots, chaps and cowboy hats makes us think of a different and better way of life. The cowboy code of honor is an image that is alive and well in our hearts and minds, and its wisdom is timeless.

Using ranch based stories, author Michael Gooch, a ranch owner, tells us how to apply cowboy wisdom to our everyday management challenges. Serving up straight forward, practical advice, the book deals with issues of dealing with conflict, strategic thinking, ethics, having fun at work, hiring and firing, building strong teams, and knowing when to run from trouble.

A unique (and fun!) approach to management training, Wingtips with Spurs is a must read whether you are new to management or a grizzled veteran.

ISBN: 1-897326-88-2   (paperback)

Also available in ebook formats. Order from your local bookseller, Amazon.com, or directly from the publisher at

http://www.mmpubs.com

A New Book by Ida Shessel

MEETING with SUCCESS

Tips and Techniques for *Great* Meetings

Are People Finding Your Meetings Unproductive and Boring?

Turn ordinary discussions into focused, energetic sessions that produce positive results.

If you are a meeting leader or a participant who is looking for ways to get more out of every meeting you lead or attend, then this book is for you. It's filled with practical tips and techniques to help you improve your meetings. You'll learn to spot the common problems and complaints that spell meeting disaster, how people who are game players can effect your meeting, fool-proof methods to motivate and inspire, and templates that show you how to achieve results. Learn to cope with annoying meeting situations, including problematic participants, and run focused, productive meetings.

ISBN: 1-897326-15-7   (paperback)

Also available in ebook formats. Order from your local bookseller, Amazon.com, or directly from the publisher at

http://www.mmpubs.com/

Your wallet is empty? And you still need to boost your team's performance?

Building team morale is difficult in these tough economic times. Author Kevin Aguanno helps you solve the team morale problem with ideas for team rewards that won't break the bank.

Learn over 100 ways you can reward your project team and individual team members for just a few dollars. Full of innovative (and cheap!) ideas. Even with the best reward ideas, rewards can fall flat if they are not suitable to the person, the organization, the situation, or the magnitude of the accomplishment. Learn the four key factors that will *maximize* the impact of your rewards, and *guarantee* delighted recipients.

101 Ways to Reward Team Members for $20 (or Less!) teaches you how to improve employee morale, improve employee motivation, improve departmental and cross-organizational teaming, maximize the benefits of your rewards and recognition programme, and avoid the common mistakes.

ISBN: 1-895186-04-8   (paperback)

Also available in ebook formats.

http://www.mmpubs.com

Agile Leadership and the Management of Change: Project Lessons from Winston Churchill and the Battle of Britain

Around the turn of the millennium, there was a British poll that asked who was the most influential person in all of Britain's history. The winner: Winston Churchill. What distinguished him were his leadership qualities: his ability to create and share a powerful vision, his ability to motivate the population in the face of tremendous fear, and his ability to get others to rally behind him and quickly turn his visions into reality. By any measure, Winston Churchill was a powerful leader.

What many don't know, however, was how Churchill used his leadership skills to restructure the British military, government, and even the British manufacturing sector to get ready for an imminent enemy invasion in early 1940.

Learn how Churchill acted as the head project manager of a massive change project that affected the daily lives of millions of people. Learn about his change management and agile management techniques and how they can be applied to today's projects.

ISBN: 9781554890354 (paperback)
ISBN: 9781554890361 (PDF ebook)

http://www.mmpubs.com/

Project Lessons from The Great Escape (Stalag Luft III)

While you might think your project plan is perfect, would you bet your life on it? In World War II, a group of 220 captured airmen did just that – they staked the lives of everyone in the camp on the success of a project to secretly build a series of tunnels out of a prison camp their captors thought was escape proof. The prisoners formally structured their work as a project, using the project organization techniques of the day. This book analyzes their efforts using modern project management methods and the nine knowledge areas of the *Guide to the Project Management Body of Knowledge* (PMBoK).

Learn from the successes and mistakes of a project where people really put their lives on the line.

ISBN: 1-895186-80-3   (paperback)

Also available in ebook formats.

http://www.mmpubs.com/escape

CPSIA information can be obtained at www.ICGtesting.com
Printed in the USA
BVOW012319161011

273728BV00004B/29/P